MANAGING EUROPE'S WAT

Managing Europe's Water Resources
Twenty-first Century Challenges

CHAD STADDON
University of the West of England, UK

Routledge
Taylor & Francis Group

LONDON AND NEW YORK

First published 2010 by Ashgate Publishing

2 Park Square, Milton Park, Abingdon, Oxon OX14 4RN
711 Third Avenue, New York, NY 10017, USA

Routledge is an imprint of the Taylor & Francis Group, an informa business

First issued in paperback 2016

British Library Cataloguing in Publication Data
Staddon, Chad.
  Managing Europe's water resources : twenty-first century
  challenges.
  1. Water-supply--Europe--Management. 2. Water quality
  management--Europe. 3. Integrated water development--
  Europe.
  I. Title
  333.9'115'094-dc22

Library of Congress Cataloging-in-Publication Data
Staddon, Chad.
  Managing Europe's water resources : twenty-first century challenges / by Chad Staddon.
     p. cm.
  Includes bibliographical references and index.
  ISBN 978-0-7546-7321-7 (hardback) -- ISBN 978-0-7546-9900-2 (ebook)
   1. Water-supply--Europe. 2. Water resources development--Europe. I. Title.

  HD1697.A5S73 2009
  333.910094--dc22

                                                              2009026219

  ISBN 13: 978-0-7546-7321-7 (hbk)
  ISBN 13: 978-1-138-25998-0 (pbk)

Contents

List of Figures, Photos and Tables

Figures

Photos

Tables

Foreword

Water was the matrix of all the world and of all its creatures. All the metals of the world all the stones and glittering rubies, shining carbuncles, crystals, gold and silver are derived from it.

(Paracelsus, Swiss physician and alchemist – 1493-1541)

Water is an elemental life force. Until recently it has been regarded as an infinite resource to meet man's needs by modern societies. Because the distribution of accessible clean drinking water across the world has not been ideal, people from ancient times have used the civil engineers to find, contain, channel and harness water to meet their needs. The story of advancing civilisation and human development are inextricably linked to water and our control of it.

But climate change coupled with an ever increasing demand for water by the world's growing population has now increased the tensions and debates over water. In 2008 the department of international development in the UK warned that two thirds of the world's population will live in water stressed countries by 2025. A coalition of international charities has now launched an appeal to bring running water to the developing world and better sanitation to 2.6 billion people. Campaigners say that action is necessary to prevent an escalation of competition for water which could destabilise communities and descend into conflict. So despite advances in modern technology and engineering the supply of an adequate and affordable supply of clean potable water throughout Europe remains a challenge. And the solutions to these many problems and challenges will have to be pan European if not global.

It is in this context that Chad Staddon has produced an important and comprehensive work on the background to, and challenges of, water management in the twentieth century.

Described by Chad Staddon himself as a general introduction to modern water management for people from all walks of life, this is an essential read for anyone interested in water management. In a series of self contained but related and linked chapters, Chad Staddon sets out a clear and comprehensive overview of the historical, social, environmental and economic drivers of the supply, demand and management of water resources in Europe.

Staddon's approach to the book, to set out in each chapter the key issues under discussion; provide useful background case studies to illustrate his arguments; and some pointers to the reader on how they might approach further learning, will mean that this work will be particularly helpful to students and teachers of water management.

At the heart of this discourse Staddon also poses some fundamental challenges to the increasing trend among European governments, which is to believe that the solutions to our many collective water problems lie in the privatisation and commercialisation of water services. The privatisation of water services in England and Wales has delivered many benefits for water consumers over the last decade. But with the advent of climate change and the continuing need for expensive investment in our water and sewerage infrastructure, new and more radical approaches to water management will be required for the next decade in which communities as well as consumers will need to be fully engaged. It is doubtful whether the market and metering between them can really be the long term answer to good water governance and the sustainability of water.

As Lyndon B. Johnson as US president reminded the world back in 1966, we are in a race:

> Either the world's water needs will be met or the inevitable result will be mass starvation and mass poverty.

This race has been going on since man first started to populate the earth.

The race continues.

Dame Yve Buckland
Chair Consumer Council for Water for England and Wales

Chapter 1

Introduction:
Water and Human Civilisation at the Beginning of the 21st Century

Introduction

It is difficult to conceive of any other element which is more central to human existence and to the human imagination than water. Biologically we are more than 2/3 composed of water and our cultures, around the world and down through the ages, have developed a complex and exalted conception of this simple but versatile molecule, two parts Hydrogen and one part Oxygen. H_2O. At its simplest we can equate water with life itself. Even a minor deficiency in water – say 5% of biophysical need – can seriously debilitate a human being. We can survive weeks without food, but only days without water (less in hotter, more arid places). And yet, as the economist Adam Smith pointed out long ago, there is the enduring paradox that while water, which is vital for life, is often considered valueless – a free good – (though this is indeed changing rapidly as we shall see in subsequent chapters), diamonds, which are biophysically useless, are highly valued. Strange priorities indeed! And Abraham Maslow, whose famous "hierarchy of needs" is well known throughout the social sciences, ranked the need for water *third*, right after "breathing" and "food"![1]

Of course many observers counter that, contra Smith, water is not valueless, but rather *invaluable* (and this distinction is at the heart of contention over the privatisation of drinking water provision in Europe and around the world explored in Chapter 6). More than two centuries ago Benjamin Franklin quipped "When the well runs dry, well shall know the value of water." As Table 1.1 on p. 18 shows a significant proportion of the world's population is approaching the point where they may not have adequate resources to supply minimum quantities of clean water (defined by the UNDP as 50 litres per person per day for drinking, cooking, cleaning and sanitation – cf. Gleick, 1999) not to mention the needs of industry and agriculture. While almost 100% of North Americans and Europeans have access to abundant clean drinking water in the home:

---

1   In fact there is a good argument for swapping food for water in the rankings since an average healthy individual can last a lot longer without food (weeks potentially) than they can without water (days only).

- only 28% of Kenyans
- only 38% of Congans
- only 69% of Mexicans

have such access (Gardiner-Outlaw, 1997; Gleick, 1999). Moreover something like 40% of the world's population does not have access to adequate sanitation. Even within the "developed" West the maintenance of access to clean water is becoming a serious problem for governments. For several years now the US state of Georgia and the city of Atlanta in particular, have been struggling to deal with long-term chronic shortages of water (Jarvie, 2007). Barcelona Spain, Melbourne Australia and many cities in the Mediterranean Basin are also facing up to mounting – and chronic – water shortages. In 2008 Barcelona was forced to import water via ocean-going tankers from other parts of Spain and from France to alleviate its shortages. Moreover not only is water distributed highly unevenly around the world, but the political economic juggernaut of privatisation of our water resources is also further eroding access to this most basic resource. Water, it turns out, is *extremely valuable*, especially if it is locally scarce and/or too expensive (which may amount to the same thing – see Chapter 6).

Even that vast preponderance of the world's water that is saline, and therefore undrinkable, is critical to human survival inasmuch as it serves irreplaceable functions of climatic, hydrological and energetic regulation at the global scale. Much current debate about climate change hinges on the changing temperature profiles of the earth's seas and oceans. Climate change science has already shown us that even minor variations in mean sea temperature may be linked with the rise in severe weather events such as hurricanes and cyclones. Through the crucial sun-driven process of evapotranspiration, the oceans themselves are ultimately the most important source of fresh water. As I write this chapter (late July 2007) parts of the UK are suffering severe flooding as a result of record 24-hour rainfalls after a lengthy dry period. Thus, the Earth's hydrosphere provides both essential resources supporting human life as well as potentially devastating natural hazards. Understanding and managing water in all its divergent guises – as resource and as hazard, as freshwater and as saltwater – is a critical challenge for the 21st century.

In this introductory chapter I sketch out some of the enduring problematics, socio-cultural, economic and political as well as strictly hydrological, surrounding the management of water resources in the 21st century. Whilst this chapter is wide-ranging and global in scope I will conclude with some reflections on specifically *European* challenges which set the tone for the remainder of the volume. First I will discuss deeper issues related to the human cultural relationship with water, before moving on to consider the relations between control over water and urbanisation, the leading anthropogenic driver of our current water crisis. This will lead to the introduction of the idea of the "engineering paradigm" – a way of thinking about water as everywhere and always a *technical* problem amenable to technical solutions. Unfortunately for us this engineering paradigm, with its emphasis on supply-side solutions (more dams, more machines, more corporate involvement) has created at least as many problems for the 21st century as it solved in the 19th and 20th centuries. In the final sections of the chapter we will explore the current (and changing) global distribution of water resources and the fundamental character of water as a "common" versus a "private" resource. Together these elements, our socio-natural conceptualisation of water, the capitalist urbanisation of water (including the "engineering paradigm") and changes to the already unequal global distribution of water (some induced by climate change), create the constitutive problematique organising this volume.

Water and the Human Imagination

Given its biophysical centrality it is hardly surprising that water exercised an early and a strong hold on sociocultural development. In this section we explore just a few expressions of this "hydro-cultural" complex from antiquity to the present. Our purpose is two-fold. First, we hope that this brief excursus will sensitise readers to the cultural importance of water and if some readers subsequently choose to develop this knowledge historiographically or anthropologically, then so much the better.[2] Second, and more importantly for the purposes of this volume, I hope to conclusively show that contemporary political economic discourses about water, about its control, ownership, disposition, etc. are very much products of a long social and cultural history. In other words I aim to destabilise the essentially technocratic "engineering paradigm" which holds that water is a purely technical problem.

The ancient pre-Socratic (7th-6th century BC) division of the cosmos into "air, water, fire, earth" was arguably logical inasmuch as it was based in immediate human experience – without microscopes, computers, electricity or automobiles surely these four physical things would have seemed fundamental, elemental and constant. But of these water exercised the strongest hold not only on the economic organisation of early societies and physical location of their principal settlements (see below), but also on their (our) collective imaginations. The Pre-Socratic philosopher Thales of Miletus (624-546 BC) believed that water was itself the most fundamental reality and founded his natural philosophy upon this principle (Lundberg, 1992). For Thales and his followers the principle task of natural philosophy was to develop explanations of the world as people experienced it (rather than as imagined in theology) and the principle elements, earth air, fire and water played a central role in this system.

Such a conception was not at all unique to the early Greeks. Indeed, the merest acquaintance with any of the world's major philosophical and religious traditions demonstrates that water is almost universally linked with pure knowledge and spirituality. As shown by the boxed quotation from *Revelation*, it is sometimes

---

2 The literature on cultural conceptions of water is rich and fascinating – see for example de Villiers (2000); Halliday (2001); Lundberg (1992) and Squatriti (2001).

socially constructed as an analogue of wisdom itself. Similarly, though certainly not a Christian, the Roman orator Seneca (d. 39 AD) declaimed that "where a spring or river flows, there should we build altars and offer sacrifices." Right up to the decline of the Empire in the 4th century AD Roman villa designs generally incorporated votive water features, usually fountains, springs or wells, such as the one at Chedworth in Gloucestershire, England. And in Confucian thought it is very important to orient oneself and one's home very specifically towards watercourses in order to benefit from its life-giving energies (the collection of principles that govern this are known as "feng shui"). And Lao Tzu (d. ca. 550BC) commented that "There is nothing softer and weaker than water, And yet there is nothing better for attacking hard and strong things. For this reason there is no substitute for it."

> "and He showed me a pure river of water of life, clear as crystal, proceeding out of the house of God and of the Lamb."
>
> Revelation, 22:1

In addition to quotations like that from *Revelation* above, other Judeo-Christian writings frequently celebrate the life giving qualities of water and near-divinity of those who make it available for human use, as in the following snippet of song from the *Apocrypha* (quoted from de Villiers, 2000, p. 71):

Spring up, Oh Well,
Sing Ye to it
Thou well dug by princes
Sunk by the nobles of the people
With the sceptre, with their staves
Out of the desert, a gift...

A different socio-religious tradition, Islam, similarly celebrates water. The *Qur'an* designates water that falls from the heavens as a sign of divine Grace and water is itself understood as the Spirit of God (*Qur'an* II, 9), a divine essence in which to bathe before praying. And the *Qur'an* is quite strict also not just about the proper keeping of wells, but also about the obligation of well-owners to provide any surplus water to those less well off – a very important form of community charity in the dry places of the Middle East and North Africa. A famous "hadith", or commentary on the life of the Prophet, relates the story of how Muhammed exhorted Uthman (one of his immediate successors) to buy a well and turn it over the community, an example followed by Muslims around the world and up to the present day.

Places inspired by Islamic culture and philosophy invariably showcase water in ponds, canals, fountains, and wells. For example at the Alhambra Palace in Granada, Spain water is an integral element in the overall design of the palace complex; it runs through buildings in artistic canals in the Court of the Myrtles and the Court of the Lions (Photo 1.1), is used in fountains and gardens (as in the nearby Generalife gardens) and is, of course, used in sanitation. Elsewhere in the Islamic world the 17th century Shah Jahanabad (known in English as the

Photo 1.1   Court of the Myrtles, Alhambra Palace, Granada, Spain
Source: Author.

"Red Fort") built in Delhi used water to provide natural air conditioning as well as decoration and architectural coherence. Similarly all mosques incorporate water fountains at or near their entrances in order to facilitate the ablutions necessary to prepare for daily prayers ("wudu" as codified in the *Qur'an* above). In fact there are also strict rules about the quality of water to be used for wudu: it must be clean and pure and not previously used for ritual washing.

Of course water has long been elemental not only to the human imagination, but also to survival and, beyond that, social order and spatial organisation.[3] It is an empirical fact that virtually all major towns and cities of the ancient world are located as much with a view to securing access to water as to access transportation and trade routes or even defence. The earliest cities of the Nile and Indus and Euphrates River valleys were ultimately dependent upon their riverine water sources for their existence no matter how their complex technologies of control and organisation seemed give them mastery over their natural environments. The Hanging Gardens of Babylon, apparently built in the 6th century BC by King Nebuchadnezzar and one of the Seven Wonders of the Ancient World, are perhaps one of the most obvious symbols of this dependence: a massive series of terraced roof gardens, each lined with layered reeds, bitumen and lead to prevent seepage to the levels below, all irrigated with water pumped (possibly using a sort of Archimedean screw) from the Euphrates River (de Villiers, 2000). Differently impressive was the Assyrian king Sennacherib's use of water, and of fire, to utterly destroy Babylon in the early 7th century BC (at about the same time the natural philosophers of Asia Minor were placing water at the centre of their science).

Eastern traditions too tended to put great importance on water and water-orientated ritual practices. In Hindu tradition, for more than 3,000 years the Ganesha Festival has involved the ritual immersion of effigies of the elephant god Ganesh (first son of Shiva and Parvati) in water, and the throwing of clay figures into lakes and rivers with the incantation "what begins with water ends with water". Indeed, the practice has become so ubiquitous in India that many religious organisations have begun urging the immersion of non-polluting idols made of paper coloured with vegetable dyes rather than the traditional plaster of Paris ones (*The Hindu*, 12 September 2004). And of course periodic immersion in the Ganges River is said to provide deep spiritual cleansing to Hindus, though in 2007 some Hindu leaders complained that the river had become too dirty for the ritual bath.

We could go on for a long time indeed enumerating the place of water in ancient views of the world, but surely the point is made: water is central to both human life and culture and flows through our social and cultural fabric as a generative force, something both mundane and sublime.[4] As Primo Levi (1995) writes in *The Periodic Table*:

---

3   See the recently published volume of essays on the history of water management in Europe edited by Squatriti (2002) for more information and case studies.

4   Newson (2008) makes the same point, noting that water is everywhere understood as both a practical necessity and as a spiritual essence.

Water is ever close to mankind, or rather to life, by the bonds of age-old familiarity and ever-present necessity so that its uniqueness is concealed under the guise of the habitual.

As a "static" substance it is central to our biophysical survival whilst as a dynamic resource, as a resource that moves and transmogrifies, it is associated with social change and transformation, and, according to French philosopher Bachelard, even death (Bachelard, 1942). As we shall see in the next chapter this paradoxical duality of water as both a "stock" and a "flow" resource has been an enduring problematic for the development of legal traditions around the world and through history as well (Caponera, 2007).

Human Settlements and Water

Possibly inspired by this deep cultural groundwater, various urban historians (Sjoberg, Mumford, Wittfogel) have pointed to the centrality of control over water in the urbanisation process.[5] Surely the development of the ancient "qanats"(a system designed to draw, store and distribute water over large areas) in the Middle East, North Africa and the Iberian Peninsula provides yet more eloquent testimony to the lengths to which we have always gone to control water; to have it where and when we need it (de Châtel, 2007). Many of these surface channels fed by underground tunnels, which can cover the land surface like a dense web, are still in use today. Versions of the qanat system, having diffused from their point of origin in the Fertile Crescent, are found as far east as China and as far west as Spain and even Latin America (where they are known as "acequias"). Of course the Greeks had been engineering aquaducts since the early 7th century BC and Pythagoras (c. 572-c. 490 BC) himself actually engineered what may have been the first tunnel through a hill to carry water into the town of Samos in the Cycladean Islands of the Aegean Sea. We could also point to the "acequias" of the Iberian Peninsula and Latin America, the canals of the Hunza of the high Karakoram, the aqueducts of the Roman Empire and other equally remarkable feats of hydro-engineering which have been successfully undertaken without recourse to modern technologies. It is well worth quoting Marq de Villiers' (2000, p. 79) description of Rome's 1st century AD water system at length:

---

5   Karl Wittfogel's (1896-1988) "Hydraulic Thesis" has received much attention in the literatures on urban and environmental history. In *Oriental Despotism: A Comparative Study of Total Power*, published in 1957, Wittfogel argued that the first centralised Chinese state was founded on and though control over water. Some see it is as overly reductionist (Blaut, 1993), whilst others prefer more nuanced and locally-specific theorisations (see, for example, Diamond (1998) and O'Tuathail (1994). Nobody seems to challenge the basic insight however that the existence of cities presupposes the existence of technologies and bureaucracies to control water.

> The Roman system, described in detail by Sextus Julius Frontinus, the supervisor
> of the Empire's waterworks, used eleven major aqueducts to bring water more
> than 40 km to the city in sinuous, curving channels that were themselves almost
> 100 km long, mostly in underground tunnels made of stone, terra cotta, and a
> variety of other materials...the system was gravity fed and flow-through...the
> surplus water was used to power the city's fountains and to flush its sewers into
> the Tiber.

Even this was eclipsed by the water supply system built by the Emperor Hadrian
for the north African city of Carthage in the second century AD, transporting
water more than 130 km into a city well-supplied with underground cisterns
and rainwater collection systems ("impluvium"). The technological complexity
of these systems was not really surpassed until the invention of the pressurised
piping systems of the 19th century using wooden pipes reinforced with steel bands
(Hauck and Novak, 1987).

Of course it was not just technology that was impelled by the need to control,
channel and manage water resources. Legal structures (statutes and juridical
processes) were also greatly exercised by the need to regulate human appropriation
(and misappropriation) of water. This issue is treated in greater detail in the next
chapter, but it is useful to make a few historical comments here. First, whilst
some might perceive a "privatising" impulse in technological developments such
as aqueducts, cisterns, and other waterworks, legal traditions reaching back to
antiquity place great importance on the legal identity of water as a *public* good.
As Caponera (2007) and others have pointed out, at least as far back as the second
millennia BC *Hammurabic Code*, water was defined as the grace of God and
therefore inalienable. Sanctions for the misappropriation of water ranged from
financial penalties all the way up to death. Second, water law, even up to the
present day, is very strongly related to the idea of natural justice, that is: to the idea
that each and every citizen (and now the environment itself) has an inalienable
right to sufficient water to live and to thrive (Gleick, 1999; Salman and McInerney-
Lankford, 2004; Wouters, 1997). This latter point will be of central moment in our
discussions of water law in Chapter 2 and water sector privatisation in Chapter 6.

Access to water was a key issue in early urbanisation. The first civilisations
grew up on riverbanks, on flood plains and in deltas because water was central
to agricultural production, urban hygiene and transportation. The seasonal
fertilisation of these areas through sediment-laden inundation allowed for
sustained crop production and the development of agriculture. The Indus, the
Mekong, the Euphrates, the Tigris and the Nile are all examples of rivers that
fostered civilisations, some of them going back more than 10,000 years. To cite
examples from Central Europe: at the spot where the Vistula River enters the Gulf
of Gdansk the city of Gdansk grew up 1,000 years ago; where the Daugava River
enters the Gulf of Riga, Riga was built more than 800 years ago, and where the
Neva River enters the Gulf of Finland St Petersburg was founded by Tsar Peter the
Great in 1703. London ("Londinium" to the Romans) was of course established on

the Thames River more than 2,000 years ago and Paris was founded on an island in the Seine, securing both access to water and defence at a stroke. These large rivers were important waterways for transportation, trade and travel and the deltas, as well as sources of water for drinking, for agricultural and for (incipient) industry.

Although the best known waterfront cities may be Venice, Amsterdam and Hong Kong, other parts of Europe have many cities with spectacular waterfronts. Stockholm is richly blessed with beaches, streams, lakes and seafronts situated as it is, where Lake Mälaren meets the Baltic Sea. Tsar Peter I ("the Great") designed St Petersburg, similarly situated in the River Neva delta, once with some 70 smaller and larger islands, with Amsterdam as his model. More recently many cities, including Paris, Prague and Krakow have rediscovered their own waterfronts as objects of aesthetic and recreation appreciation. Since 2002 municipal authorities in Paris have sponsored no less than three summertime "Paris Plages" along the previously largely unappealing Seine River. Further afield American and Canadian cities such as Baltimore and Vancouver have redeveloped their waterfronts as spaces of recreation, leisure and 21st century capitalist urbanisation (Harvey, 1990). In Vancouver local activists had to fight hard to maintain public access to a small patch of beachfront access in the poorer downtown eastside neighbourhood.

It is obvious that water was appreciated for both its *function* as well as its *form*. Rivers were used for potable water and for transport, but fishing, energy generation and defence considerations were important too. Less remarked upon by commentators on the current water crisis is the important use of water in waste removal (sewage). After all one logical consequence of water use is the need to get rid of it in roughly equal amounts as wastewater. The average three person household in the UK requires about 420 litres of clean water each and every day, but only a small fraction of it is used "consumptively", that is to say permanently. The remainder is returned to the waterworks system as wastewater, once it has been used in showers, washing machines and, of course, toilets (approximately 1/3 of this input water is flushed down the loo!). A sewage treatment system with technological solutions is employed to remove the disposal away from the city. The rainwater runs underground in sewers to be disposed while it is essential to notice it as a resource to be recycled and re-used. Recycled wastewater can be used in the WC or for gardening. In 1762, Nicola Salvi, the architect of Rome's spectacular Trevi Fountain, intended it to symbolise the water cycle. He wrote:

> The sea is, so to speak, a perceptual source which has the power to diffuse various parts of itself, symbolised by the Tritons and the sea Nymphs, who go forth to give necessary substance to living matter for the productivity and conservation of new forms of life, and this we can see. But after all this function has been served, these parts return in a perceptual cycle to take on new spirit and a new strength from the whole, that is to say from the sea itself (quoted in Moore, 1994).

At earlier times the purification of wastewater was simply neglected, resulting in terribly polluted and foul smelling rivers and seasides. This is still the situation in some places. In the River Neva the water leaving the city contains considerable amounts of untreated wastewater and Riga is still struggling with its sewage system to avoid direct channelling of wastewater to the river. Much runoff from streets and other paved surfaces is going directly to the closest surface water source, loaded often with petroleum by-products, etc. (see Chapter 4).

In the UK in the 19th century engineering paradigm led also to a distinctive and enduring "hydro-architecture" (towers, pumping stations, dams, etc.). The great dams of mid-Wales, the Peak District and other parts of the UK, the associated elements of Joseph Bazalgette's sewage system for London all manifest that quintessential Victorian desire to make the functional visually appealing, often with reference to neo-baroque romanticist design elements such as crenellations, turrets, etc. (e.g. Vrynwy reservoir straining tower, Figure 1.2). These edifices also rank as some of the most important technological structures of their age (cf. Halliday, 2001). In the main building at Papplewick, Nottinghamshire for example, built in 1846 to augment the growing industrial city of Nottingham's water supply, there are two massive 140 hp steam-driven pumping engines, thought to be among the last built by James Watt and Company which until 1969 lifted water from a 200-foot deep well dug into the underlying sandstone and pumped the water into the reservoir that supplied Nottingham. Papplewick is of course not unique and England and Wales are liberally furnished with other examples of this Victorian flourishing of hydroengineering.

As impressive as such technological achievements were, we need also to note that not only is the idea of the "control *of* water" central, but so too is the reality of "water *as* a control", for human societies have found that settlements cannot prosper just anywhere, and that our technological ingenuity in moving, storing, and controlling water has limits imposed by the realities of relative geographical scarcity (and the occasional sudden overabundance of water, as in July 2007 in Gloucestershire, England). From the ancient to the early modern period cities and major settlements tended to be located on or very near to water sources – their builders did not exhibit today's hubris that technology could conquer any and all problems of water storage and transport (how else can we justify cities of millions of people in the more arid parts of the American southwest?). The natural location and movement of water in the landscape was therefore as much a control on human development as was human control of water. Even the achievements of ancient Roman and Chinese water engineering were largely restricted to the local and regional scales and were thus relatively modest to contemporary eyes. This ecological balance remained, arguably, unbroken until that point in the 18th or 19th century when water itself, that basic element of human life and society, was drawn into the capitalist commodity economy and thus fundamentally transformed

Photo 1.2   Vyrnwy Tower, Mid-Wales

Source: Reprinted with permission from Severn Trent Water.

as a "socio-natural" entity (Swyngedouw, 2004).[6] I will return to the elaboration of this point in Chapters 2 and 6.

The remainder of this chapter introduces students to some of the key issues of global modern water management. Starting with a brief consideration of the global geography of water scarcity we turn to the introduction of some key terms and concepts of contemporary significance. For example, the question of water's status as a "common" or a "private" good is broached,

> *"One may not doubt that, somehow, good shall come of water and of mud, and sure, the reverent eye must see a purpose in liquidity."*
> Rupert Brooke

as is the idea that, until recently, the modern hydro-imagination has been dominated by a conviction that technology can solve all problems of water quantity and quality. This "Engineering Paradigm" has bequeathed to the 21st century some remarkable waterworks like the Aswan High Dam or London Metropolitan water system, but it has also helped to create serious problems which must now be addressed. The channelisation of rivers for example has, by all accounts, greatly increased our susceptibility to disastrous flood events, and the blind faith in dam-building has fundamentally compromised the fertility of soils across the world as millions of tonnes of nutrient-rich sediments are now impounded behind dam walls.

The Engineering Paradigm and Water Management in the 21st Century

Humans have long thought that they could engineer themselves out of virtually any difficulty, and perhaps nowhere has that ill-fated logic been more manifest than in the area of water management. Elsewhere in this volume (especially Chapters 4, 5 and 7) I discuss some of the social, environmental and health impacts of ill-conceived water developments. Of course technical know-how could not have enacted such remarkable feats of hydro-engineering (whatever your opinion of large dams from social or environmental perspectives) as China's Three Gorges project,[7] or even the Roman aquaducts of southern Europe in the absence of strong political will. And indeed the story of water management in the 19th and 20th centuries is the story of the convergence of political and technical programmes around a modernist ideal of "progress" and "development" that made such projects not only conceivable, but imperative. Still, it is important not to draw the lesson from this critique that political and technical impulses ought now to be abandoned in favour of something else. Rather, it is hoped that through a careful and critical

---

6   Similarly Malcolm Newson (2008) suggests that we have now left behind the Holocene epoch and entered the "anthropocene".

7   Construction of the project, designed to generate more than 18 million kwh of electricity and provide irrigation water, began in 1994 and initial filling of the storage reservoir began on 1 June 2003. As of summer 2008 the TGP was generating 10 terawatt hours of electricity/month.

examination of past approaches to water engineering and management it may be possible to retain the best of the engineering paradigm and conjoin it with a new political, and policy, will to manage water resources more carefully and with a view to the sustainable and socially equitable hydrological development of all European nations.

To some, particularly those beginning their professional lives in the 1990s and 2000s, it may seem incredible that water managers once truly believed that all problems of water supply, power generations, irrigation, economic development, etc. could be solved through building bigger, more expensive, more technically ambitious projects. Yet until recently this sort of naked technocentrism – an extreme expression of the engineering paradigm – was more the rule than the exception and water projects in particular led many to rapturous exhortations such as:

> What a stupendous, magnificent work – a work which only that nation can take up which has faith and boldness!…it has become the symbol of a nation's will to march forward with strength, determination and courage…as I walked around the [dam] site I thought that these days it is the biggest temple and Mosque and gurdwara where man works for mankind. Which place can be greater than this, this Bakra-Nangwal, where thousands of men have worked, have shed their blood and sweat and have laid down their lives as well? Where can be a greater and holier place than this? (Pandit Nehru, first PM of India, 1954).

Such a view betrays not just a hubristic love of technology for its own sake, but also a strong fear or loathing of nature itself. As recently as 1985 the head of a large Canadian dam engineering firm declared: "In my view nature is awful and what we do is to cure it", a view which echoes earlier declamations by Floyd Dominy of the US Bureau of Reclamation: "The unregulated Colorado [river] was a son of a bitch. It wasn't any good. It was either in flood or in trickle" (Reisner, 1993).

One of the best examples of the apotheosis of this modernist paradigm of water management is provided by the 20th century history of the Tennessee Valley Authority (TVA) in the US (Chandler, 1984). Created by President Franklin Delano Roosevelt in 1933, the TVA was intended as a key plank in FDR's national "New Deal" programme to lift the US out of the depths of the Great Depression. The TVA was, like many of today's QUANGOs, a vastly powerful organisation with feet planted firmly in both the public and private sectors. Run like a corporation, albeit one with its hands deep in the public purse, it had within 20 years of establishment built more than 20 new dams and reservoirs and had an installed hydroelectric capacity of more than 4 terawatts. Even today the TVA has a workforce of over 24,000 people and is the major supplier of electricity to over 100 Appalachian municipalities.

In this connection between hydrodevelopment and national regeneration the TVA was, if spectacular, then only one of many such expressions. In Europe there are many examples of similar corporatist thinking in big water development projects. As Swyngedouw (1999, p. 450) points out in his case study of 20th century Spain:

> Under Franco, the great expansion of hydraulic infrastructures reshaped the
> hydraulic geography of Spain in fundamental ways.

Indeed, the fascists in Spain certainly understood water development projects in
much the same way as Pandit Nehru understood dam projects in his country; as
material expressions of the regeneration of national spirit and greatness (a policy
called *regeneracionismo* in Spain in the 1950s and 1960s). Similar projects were
undertaken throughout Europe, including the Soviet-dominated countries of
Eastern Europe (Staddon, 1998). The logic of equating big hydro-development
with national (re)organisation is indeed compelling; such projects, particularly the
really big ones, require a fusion of the interests of capital, society and government
towards the ends of providing water for irrigation, drinking, industry and power
which themselves are powerful symbols of national coming together and as such
can justify even authoritarian government. As the Spanish hydro-engineer Costa
put it in 1892: "to irrigate is to govern" (in Swyngedouw, 1999, p. 456).

   Though the TVA, *regeneracionismo* and allied social movements in other
countries continue to manage the now completely de-natured river systems of
Europe and America, it is nevertheless true that the last two decades have seen a sea-
change in thinking about the best way of managing water resources for sustainable
development. In the first place, the old modernist pact between the "politics of
progress" and technology for its own sake has been largely discredited. Partly this
decomposition has come from external attack, by environmental groups decrying
the high cost to the environment of such developments, and partly this has been the
result of political retrenchment away from activist roles in the development process.
Engineers too now seem far less prone to the sorts of hyperbolic statements cited
above. Perhaps we could say that the modernist pact between politics, technology
and corporatist development (big companies backed by big government) has
given way to a new pact between these three refracted through the now ascendant
ideology of "sustainable development". As classically stated in the 1987 report
"Our Common Future" the agenda of sustainable development is neither anti-
political nor is it anti-technological. Rather, it is orientated towards a much more
humane and environmentally sensitive interpretation of the *manner* in which
development ought to be pursued. The Canadian Water Resources Association
(CWRA), for example, articulated its interpretation of sustainable development of
water resources in terms of achieving a balance between ecological integrity and
social equity (Mitchell, 1997). Note though that this does not in any sense preclude
big water projects of the sort that generated the popular protests that helped spur
the idea of sustainable development in the first place – this fact has led to a good
deal of criticism from environmental sceptics.

   One of the new approaches emerging from new thinking about sustainable
water management is called "Integrated Water Resources Management" and it
involves state, corporate and environmental groups in a collective project of water
management, thought through in a much more holistic way. Initially attempted in
the 1970s, the key objectives of integrated management are:

- Maximise the efficiency of water use
- Maximise water availability by minimising stock degradation
- Optimise allocation to competing users
- Minimise waste of water

Though these objectives may seem obvious, they require water managers to undertake many more tasks than previously, including demand as well as supply management, and in consultation with many more "stakeholders" than before. For example it is necessary to have a common comparable database of water resource, withdrawal, consumption and quality upon which to base regional and interregional decision-making – something which has not existed in most countries prior to the 1990s (Calder, 2005; McDonald and Kay, 1988). Consequently, nations and regions which have begun to go the route of integrated management have had to face up to the need to undertake significant institutional reform, the restructuring of the water supply industry, and public education about water management possibilities. To date there are relatively few good examples of functioning systems of Integrated Water Resource Management, though useful starts appear to have been made in Canada (Pearce et al., 1985) and in the Murray-Darling River Basin of south Australia (OECD, 1998). Agenda 21, coming out the 1992 UNCED conference in Rio de Janeiro also provides a useful statement of integrated water management as a key component of the overall objective of sustainable development. In the 27 member European Union a strong start has been made in this direction with the passage of the Water Framework Directive in late 2000.

An "integrated" approach must take proper account of the fact that the hydrological cycle combines natural and anthropogenic (human-induced) processes – it is in fact a "hydro-social" cycle. As Figure 1.1 clearly illustrates, there is little sense in any longer holding to the "common sense" idea that there is a "natural" water cycle which is intersected at a large number of identifiable points by human activities. Rather, the basic idea underlying the integrated approach is that the hydrological cycle, for most parts of the globe, has now become fully anthropogenic. Thus we can see that what "begins" with evaporation from the sea takes water through a series of natural-physical processes, and also through anthropogenic ones in a complex set of interacting feedback loops. Indeed the figure no doubt presents a vastly simplified picture and it is easy to see that there must be many more layers of feedback loops and interaction arrows in even a modest water supply area. There are two central implications of this movement towards "integration" in water resources management. First, managing water has become much more closely linked to managing peoples, societies, economies and polities. Second, consultative approaches are increasingly called for, rather than the old reliance on the "experts" (usually engineers or their political cheerleaders) to get things right.

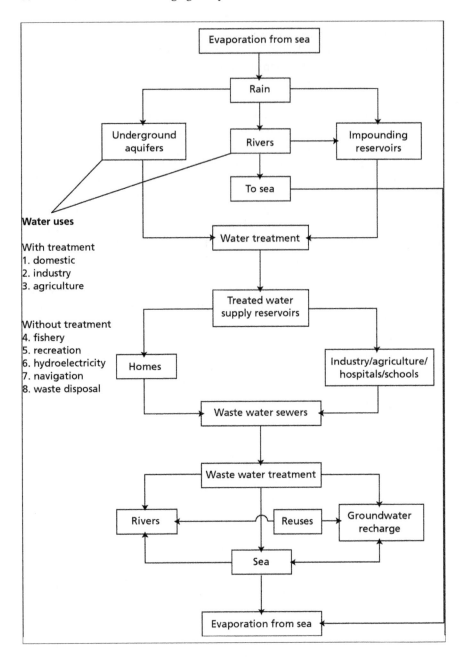

**Figure 1.1   Integrated Hydrological Cycle (including water supply, use and
               wastewater cycles)**

Source: Author.

The Pale Blue Dot: Water Scarcity at the Beginning of the 21st Century

From space the Earth appears quite dominated by water: over 70% of the earth's surface is covered by water; water in lakes, rivers, icecaps, aquifers, and especially oceans. From this perspective might have thought that the planet would have been called "Water" rather than "Earth"!

But this apparent abundance of water betrays real scarcity, for only about 3% of the earth's total water resources are freshwater supplies – the vast majority being locked away in saline environments; oceans and estuaries (Shiklomanov and Rodda, 2004). Moreover, only about 10% of that 3% is actually available for easy exploitation, the remainder being locked up in other flora or fauna, in underground aquifers (about 30%), glaciers (about 65% of total) or the atmosphere – implying that humanity is surviving on approximately 0.3% of total world water supplies; a thin trickle of life-sustaining liquid indeed! But things are not as bad as they might seem, with over 100,000 km³ (freshwater) precipitation deposited annually and only about 500 km³ used consumptively there would appear to be ample supplies, at least at the global level. Added to the 105,000 km³ or so of water held in lakes and rivers and this begins to seem not merely an abundance but a super-abundance.[8] Overall human societies are only using about 1% of total

Photo 1.3   The Earth from space

Source: NASA.

---

8   Some experts (e.g. Rogers, 2008) suggest that the minimum amount of water required by each and every person for drinking, hygiene and growing food is about 1,000 m³/yr. By this standard, the earth should be able to provide water for more than ten times the current world population of just under seven billion.

annual renewable resources and even pessimistic estimates suggest that total global availability of freshwater could remain above 5000 m³ per person per year.

The problem however is one of *distribution*, for these annual rains are distributed *highly unevenly* geographically; some regions have far in excess of the 5,000 m³ identified above, some very much less (Gleick, 1993; Postel, 1992). We know, for example, that some regions receive almost no annual precipitation, whilst others receive a great deal; so too seasonal distributions vary widely, from the year round patterns of the Marine West Coast climates to the seasonal monsoons of south and southeast Asia. The UK for example, during the 1990s and early 2000s experienced serious water shortages in some regions (and floods in others) not because annual rainfall was down, but because the seasonal distribution of that rainfall had become quite uneven. Even the ensuing (and current) relatively wet period has been complicated by significant regional disparities, as discussed further in Chapter 8.

Table 1.1　　Freshwater resources, 2005

	Resources	**Withdrawal**	**Balance**
Iceland	582192	543	581649
Canada	91419	1494	89925
Russian Fed	31653	527	31126
Indonesia	12749	391	12358
US	10333	1682	8651
Netherlands	5608	500	5108
UK	2474	163	2311
China	2206	494	1712
India	1754	635	1119
South Africa	1106	348	758
Algeria	750	161	589
Jordan	157	202	-45
Israel	255	338	-83
Saudi Arabia	337	518	-181
Kuwait	8	198	-190
Egypt	794	1013	-219
Libya	106	919	-813
World Average	8549	633	7916

Source: World Resources Institute (internal water resources and withdrawals in cubic metres per capita per year, 2005).

Table 1.1 depicts some of the gross disparities in water's distribution across the globe. Overall water stock resources for a selection of countries are indicated in the first column, current withdrawals in the second and the resulting "balance"[9] in the third. Those countries with a water balance in excess of 2,000 m³/capita are considered "water rich" whilst those with less than 1,000 m³/capita are considered "water short". Middle range countries, like India and China have 1,000-2,000 m³/capita and are considered "water scarce". It is immediately obvious, from the data presented, that not only are there countries of clear abundance (Canada, Iceland) and countries of worrying shortage (Egypt, Libya, Kuwait), but also *regions* of abundance (North America) and *regions* of shortage (North Africa and the Middle East). Yet these data present at best a partial picture, since even physical hydrological abundance does not guarantee that everyone has equal access to the water they need. Nor do these statistics take account of the likely impacts of climate change, particularly in terms of making much of even the relative abundance enjoyed by the UK unavailable by concentrating it into higher frequency storm events. Both of these issues are discussed later in this volume (particularly in Chapter 8).

In the UK the average per capita *direct* consumption of water is approximately 150 litres per person per day or the equivalent of 55 cubic metres per year which seems very favourable (compared with the water availability figures presented above) until one realises that this figure *does not* include the use of water for food production (agriculture), industrial processes (e.g. energy generation) or transportation of goods and people. Nor do these figures tell us much about the "virtual water" imported into the UK in the form of food and other products – often from water-stressed regions (Allan, 2000, Berrittella, 2007; Hoekstra and Chapagain, 2005; Renault and Zimmer 2003) or indeed the astonishing growth of bottled water markets around the world (Royte, 2008). Moreover, these figures do not tell us much about regional variations in water availability *within* nations, which are often as striking as those *between* nations presented Table 1.1. For example, in the UK total precipitation varies by as much as a factor of 4 between the wettest parts of the Western Highlands of Scotland (1,800 mm/year) and the driest parts of East Anglia and Kent (450 mm/year) in the east and southeast of England. These facts and their implications for water management will be discussed in greater detail in Chapters 3, 4 and 8.

Note that a number of countries actually have a *negative* national water balance, a fact that implies that the shortfall is being made up through the over-pumping of groundwater, international water transfers or, much less frequently, through the production of fresh water from sea water through complex and expensive desalination processes. Libya's "Great Man-made River" project, initiated in the 1980s presents a classic case of unsustainable over-pumping of fossil water resources. Over the past generation Libyan water engineers have constructed a network of over 1,000 wells, some over a half a kilometre deep, to extract fossil waters from underneath the Sahara Desert and directing them through aqueducts

---

9   In Chapter 3 we will develop this idea of "water balance" further through the technical tool of assessing regional water balances.

towards the much more heavily populated coastal zone. International water transfers are becoming an increasingly political issue with many countries heavily dependent on water "from abroad" to support their populations (e.g. Israel and Egypt from Table 1.1). With respect to the latter, some commentators laud water desalination as the "key" to solving regional water shortages, especially in the Middle East, but the expense of the most common reverse osmosis systems has put them beyond the reach of all but the richest oil-economies of the Persian Gulf, Europe and North America. Even so there are something like 7,500 desalination plants, many of them tiny, in operation around the world today (de Villiers, 2000). The economics of desalination are highly contested, with estimates of cost ranging from US$1-5 per litre of drinking water created, not including the energy demands and pollution impacts "embodied" in the equipment itself.

A clean adequate supply of fresh water is critical to sustain terrestrial life, including human life. Yet as the table shows, a large proportion of the world's population does not have sufficient access. In fact, 80 countries worldwide, with 40% of the world's population are experiencing chronic or periodic water shortages (Gleick et al., 1999). The 1977 Mar del Plata Treaty, ratified by over 100 countries, supposedly guarantees an adequate supply of clean water to every person on earth. Notwithstanding the tremendous efforts and investments in water supply provision during the "International Decade for Water and Sanitation" in the 1980s, unfortunately there were as many people without access to potable water in 1990 as there were in 1980. Population Action International estimates that between 2 and 3.5 *billion* people – as much as 50% of global population – will be living in regions of water stress or shortage by the year 2050 (Engleman, 2002; Engleman and Leroy, 1993). In this context the United Nations recently declared the decade 2005-2015 the "Water for Life" decade. These legal-policy issues are taken up in more detail in the next chapter.

While virtually 100% of North Americans and Europeans have access to abundant clean drinking water in the home fewer than a third of Africans and two thirds of Central and South Americans do. Swyngedouw (2004, p. 8) further suggests that perhaps as many as 50% of "urban residents of the developing world's megacities have no easy access to reasonably clean and affordable water." Even within the "developed" West the maintenance of access to clean water is becoming a serious problem for governments. Generally this is not just because of a lack of absolute resources, but because of the pollution of large proportions of available fresh water by decades of ill-considered industrial development. As we shall see in the next chapter, the pollution threat to Europe's principal waterways led directly to the first EU-brokered water agreements for the Elbe and Rhine Rivers in the 1970s. In fact some of the first concerted attempts at pan-European action involved management of the Rhine River in the 19th century. More recently, in the 1990s, both the EU and the Council of Europe have got behind attempts by riparian nations to develop an integrated watershed approach to managing the Danube River Basin. And in 2000 the EU's "Water Framework Directive" prescribed integrated catchment management as the model for all 27 EU member countries (we discuss the WFD in greater detail in Chapter 3).

Of course it can be argued that some countries' citizens are more efficient with their water use, others more profligate. Whilst the average American uses 400 litres/day, the average Briton uses 140-150 litres/day and both seem exorbitant water users compared to the average Indian who uses only 40 litres/day. And, scandalously, it remains true that at the beginning of the 21st century peasant farmers are still being turned off their land to make way for more water wasting first world leisure facilities such as golf courses and beach resorts. Golf courses in arid regions can consume up to a quarter million litres of water each and every day! And it is not just golf courses that are draining away scarce global water resources: raising winter vegetables and flowers for Western markets also results in the export of what Tony Allan has referred to as "virtual water" (Allan, 2001; Chapagain, 2006).

The simple fact of the matter is that water-rich countries, virtually all of them already well developed economically, are among the *least* efficient in water use. As studies of the cultures of water use in Europe have shown (Medd and Shove, 2004) European levels are rising towards North American levels as Europeans seek to emulate the (water intensive) good life. Not only are absolute rates of water use high, as noted above, but the amount used "consumptively", that is to say water that is not returned to the integrated hydrological cycle, is also too high. Moreover, it seems scarcely creditable but it is true that most Western water systems tolerate water losses, due to damaged or poorly maintained water infrastructure, evaporation, etc., of as much as 50% of treated potable volumes (a recurrent scandal in the privatised water sector of England and Wales – see Chapter 6).

Contrary to the implications of the tables above, it is not always the case that high water withdrawals indicate wastage of water. After all if much of that water finds its way back into the integrated hydrological cycle, then it may be that high withdrawals are balanced off by relatively low consumption (that is, by correspondingly high returns). However, in countries like Canada, Sweden and the US a great deal of water is effectively destroyed through industrial pollution, especially in water profligate industries like pulp and paper-making and metallurgy. There is therefore a strong argument for rethinking not just total withdrawals, but also total consumption patterns.

Water: Commons or Commodity?

Before we go much further it is worth thinking a little more about the *philosophical* underpinnings of contemporary water management strategies. We know that water is "obviously" a "common resource"; that is, it is common to us and a public good of sorts, but what does this mean for management of water quality? How can private individuals and/or groups be held responsible for their impacts on a common good? How can we optimally manage a good that belongs at once to everyone and to no-one? Or one that is, as Adam Smith noted, both *value-less* and *invaluable*. Can there be said to be an international "right" to water, as some suggest (Salman and McInerney-Lankford, 2004; Wouters, 1997)? Water is a particularly slippery commodity in that is not only a "common", but it is also one that moves across the landscape, occasionally changing its physical properties (freezing, evaporating, condensing, etc.) and which can also quickly change from resource to hazard, as the English floods of June and July 2007 so forcefully reminded us.

Fundamentally there are two ways of thinking about water as an object of management. Proceeding from the above one can try to construct management systems

Table 1.2   Water scarcity in 1995 and 2025

Country	Population 1995 (millions)	Water Per Capita 1995 (m³)	Population 2025 (millions)	Water Per Capita 2025 (m³)
Algeria	28.1	527	47.3	313
Bahrain	0.6	161	0.9	104
Barbados	0.3	192	0.3	169
Burundi	6.1	594	12.3	292
Cyprus	0.7	1,208	1	947
Egypt	62.1	936	95.8	607
Ethiopia	56.4	1,950	136.3	807
Haiti	7.1	1,544	12.5	879
Iran	68.4	1,719	128.3	916
Israel	5.5	389	8	270
Jordan	5.4	318	11.9	144
Kenya	27.2	1,112	50.2	602
Kuwait	1.7	95	2.9	55
Libya	5.4	111	12.9	47
Malta	0.4	82	0.4	71
Morocco	26.5	1,131	39.9	751
Oman	2.2	874	6.5	295
Rwanda	5.2	1,215	13	485
Saudia Arabia	18.3	249	42.4	107
Singapore	3.3	180	4.2	142
Somalia	9.5	1422	23.7	570
South Africa	41.5	1206	71.6	698
UAE	2.2	902	3.3	604
Yemen	15	346	36.6	131

Source: Gardner-Outlaw and Engelman, *Sustaining Water, Easing Scarcity: A Second Update*, Washington, D.C., Population Action International, 1997. Gardner-Outlaw and Engelman base their calculations on UN Population Division population estimates.

based on the above view of water as a "commons". This, as we shall see, has inspired a powerful body of research into the institutional structures necessary for sustainable management of common property associated with the institutional economist Eleanor Ostrom (winner of the 2009 Nobel prize for Economics). The other, diametrically opposed, viewpoint contends that notwithstanding its particular characteristics water is just another factor in our modern economic system based on private property and free markets. According to this way of thinking the challenge for water managers is to reposition water as, as the 1992 Dublin Statement has it, "a resource with an economic value in all of its competing uses" (see also Chapter 2; Anderson and Snyder, 1999). While it may be that philosophically views of water *as a common resource* and water *as a market resource* are incompatible, the contemporary reality is that we live in a world where both views are held at least to some degree and are reflected in the current structure of water services provision in Europe. It is therefore worth briefly outlining the two approaches. Indeed currently water management law and institutions at all spatial scales have emerged out of this crucible.

Water as a Common Resource: An "Ostrom" Approach

A *common pool resource* is one which, like river or lake, is physically structured in such a way that it is difficult to keep alternative potential users from enjoying it (Ostrom, 1990; 1999; 2000). It is thus available to all to enjoy (or to abuse). Examples of common pool resources include river systems, fishing grounds, the atmosphere, unfenced pastures and forests. The long term consequences of open access use of a common resource are a matter of some concern. A pasture, for instance, can allow for a certain amount of grazing to occur each year without the core resource being harmed. In the case of excessive grazing, however, the pasture may become progressively more prone to erosion and eventually yield less benefit to its users. Common-pool resources are often subject to the problems of congestion, overuse, pollution, and potential destruction unless harvesting or use limits are devised and enforced. But how can this be done with a resource that is, by definition, open to all to enjoy? This is the problem identified in Garret Hardin's famous 1968 essay "The Tragedy of the Commons".[10] One way to limit use of environmental resources, implicit in Hardin's critique, is to make them subject to private ownership – in which case the private owner controls access to the resource and transgressions of these exclusive use rights are sanctioned by laws (of trespass, theft and tort) and actively protected by various agencies of the state including the police and the courts. But this is neither the only nor the most common way of managing access to common environmental resources.

Someone who has given considerable thought to the question of *common pool resource* management is Eleanor Ostrom. Unlike Hardin and his followers,

---

10   In fact only part of this seminal article treated common property regimes (and that only with reference to an early 19th century theorist!). The real target of this essay was what he called the "freedom to breed", and thereby overpopulation.

Ostrom begins with the observation that there are a number of enduring common property regimes around the world and she and her colleagues seek to extrapolate their common lessons for the design of new *common pool resource* management systems. Through her empirical research on "self-organized, common-pool resource regimes (CPRs)", she has come up with a list of attributes that increase the likelihood that productive governance arrangements will be formed and will perform better than either markets or states.[11] Analysing the design of long-enduring CPR institutions, Ostrom (1990; 1999; 2000) identified eight design principles which are prerequisites for a stable CPR arrangement:

- clearly defined boundaries marking what is common from what is not;
- congruence between appropriation and provision rules and local conditions;
- collective-choice arrangements allowing for the participation of most of the appropriators in the decision making process;
- effective monitoring by monitors who are part of or accountable to the appropriators – ie. they have legitimacy;
- graduated sanctions for appropriators who do not respect community rules;
- conflict-resolution mechanisms which are cheap and easy of access;
- minimal recognition by the state of rights to organise;
- in the case of larger CPRs: organisation in the form of multiple layers of nested enterprises, with small, local CPRs at their bases.

Common property regimes typically function at a local level to prevent the overexploitation of a resource system from which fringe units can be extracted and enact all of the above through a mixture of institutionalised (and formalised) structures and local (often traditional) practices.

We have already observed that there are many examples of the collective management of water resources including the Middle Eastern "qanat" system, Spanish and Latin American "acequias" and the long-term regulation of access to water (and pastures, etc.) under English Common Law traditions. In the case of the acequias of southern Spain, local "water mayors" are appointed to manage the allocation of water resources and mediate conflicts that may arise. Under English common law water itself cannot be privately owned and indeed users (who are licenced by the state) are prohibited, whatever else their abstraction licences may say, from preventing access to, or compromising the health of the water body. All of these systems seem to combine most (if not all) of the attributes identified above, but are usually also inbricated within tightly-knit locally-orientated social systems where reward as well as sanction can be easily personalised. In his book

---

11   It is not possible to provide a detailed introduction to Ostrom or to *Common Pool Resource* theory here. Interested students should refer to the Bibliography at the end of the book for further reading.

Collapse Jared Diamond (2005) talks about one such case where a local water commissioner (in America) not only had the power of his (institutionalised) office behind him, but also was himself so highly respected that few ever challenged his adjudications. A challenge for 21st century Europe is to come up with institutional structures for common property resource management in the context of locally disaggregated, and indeed globalised, political economic systems.

One relatively simple challenge has been to develop a reliable and robust system for managing *information* about European water resources. Here, the issue involves "Europeanising" the flow of quantitative and qualitative information by eliminating national peculiarities in data gathering and management and thereby encouraging the emergence of Europe-wide standards of "best practice". As we shall see in Chapter 3 this has been achieved through the establishment of institutions such as the European Environment Agency and its periodic environmental reports starting with the Dobris Assessment of 1995. Since the mid-1990s national governments have also moved to harmonise their information systems regarding water quantity and quality (as well as other environmental resources). Such harmonisation should, of course, also remain sufficiently supple to accommodate regional and local differences.

In the broader context of EU lawmaking, common property institution-making would have to take due account of the well-established principles of "subsidiarity" and "proportionality". Respectively this means that actual implementation of new rules and regulations should be as close to the point of application/impact as possible and that their potential benefits should not be outweighed by their costs. Thus, effective CPR institutions should not deal with an issue or make decisions about a policy that could be handled more effectively or more legitimately at a lower level of aggregation, i.e. at the level of Member States or their sub-national units. Similarly no body should take a decision whose effects in financial cost, social status or political influence (especially for those not participating in it) is disproportionate either to the expectations inherent in their original charter or general standards of fairness in society. Of course it is also necessary that decision-making and dispute resolution mechanisms must be transparent in order to avoid extra-judicial or extra-regional appeals. In other words the important thing here is not that all participants will always get what they want from the system, but that they should feel sufficiently powerful within it that they do not elect to make appeals to external mechanisms for allocation or dispute resolution.

In Ostrom's terms, adherence to these factors or principles could lead to the formation of resource management units that have the best chance of optimising outcomes for all participants. The *Water Framework Directive* (see Chapter 3) could be seen as one way forward in the construction of a common property regime for water resource management at the European scale.

Neoliberalism and Water as a "Private" Good

Within the world of water management, the institutions-based approach is to some extent countered by the view that market-based allocation should have a role, perhaps even the primary role. Indeed, since the 1980s the neoliberal faith in markets as resources allocation mechanisms has been growing, leading, in the UK, to the privatisation of water supply and sewerage services in 1989 (this will be discussed further in Chapter 6). Moreover at the European scale a market for the trading of air pollution rights has been established since 2001 and indications are that a similar system could be applied to water pollution before terribly long.[12] There is a powerful, and currently quite effective, cabal of academics, governments and other institutions which seek to irrevocably "privatise" water utilising arguments from so-called "free market environmentalism" (cf. Lomborg 1998; Segerfeldt, 2005; Snyder and Anderson, 1999).

Such mechanisms are also used to regulate and manage the pollution of water resources, but not in any direct sense. Economic means include a system of charges for discharging pollution into the environment, including water, and fines for excessive concentrations or loads of discharged pollutants. Markets for water pollution rights do not yet exist but there is a de facto "market" in pollution fines: polluters are currently fined by national governments for pollution over and above limits set in water discharge permits or codified in national norms. Ideally the charge and fine system should be designed to ensure (together with charges for water intake and other water resource uses) the financial self-sufficiency of water management and the maintenance of water quality at an acceptable standard the full economic cost of water pollution. For the system to operate correctly and efficiently the law must be absolutely observed and enforced transparently and fairly (again raising issues about *institutional design* discussed in the last section).

Among the economic instruments applied in water quality protection we may list water permit trading and the pollution charges (fines) collected for those pollutants which pose the greatest hazard to water (see also Chapter 4). For example, in Poland (as elsewhere in the European Union) charges for wastewater disposal are computed for the following substances: Biological Oxygen Demand (BOD_5), Total Suspended Solids (TSS), total chloride and sulphate ions, volatile phenols and total heavy metal ions (see Chapter 4 and Appendix 1: Glossary of Technical Terms). Wastewater disposal charges are additionally differentiated depending on the source (the highest unit charges are applied to chemical industry, fuel-power generating, metallurgic, electric machine and light industries; the lowest are applied to municipal economy) and the imputed impact on the environment (i.e. one pays more for threatening more sensitive environments, which may well have special protection designations as well). Further differentiation results from the type of recipient (higher charges are collected for pollution discharged into

---

12   Tradeable pollution rights were originally established in the US in the 1990s for sulphur dioxide, a primary component of acid rain, as a result of the 1990 *Clean Air Act*.

natural and artificial lakes) and the region of the country (higher charges in well developed areas). Fines are collected for discharge of effluents which do not meet the requirements set forth in the water permit and for discharges without such a permit.

In the UK in 2006 there were 108 serious (category 1) water pollution incidents reported by the Environment Agency, £2.7 million in fines were assessed (for all pollution including water, air and land) and 28 company directors prosecuted (Environment Agency *Annual Report* 2006). Fully 25% of these water pollution incidents were related to sewerage operations, including sewerage system overflows. Fines are also assessed by the Drinking Water Inspectorate for breaches of quality standards in supplied drinking water in England and Wales. According to official data, between 2000 and 2008 there were 17 formal prosecutions and 16 "cautions" for water quality lapses, approximately two DWI actions per year (DWI 2007 Annual Report [on-line]). In the single prosecution achieved in March 2004, and which is fairly typical of such cases, Anglian Water admitted liability for providing water unfit for human consumption as a result of foul taste and odour. This affected only 30 properties and, ironically, occurred as a result of planned maintenance of the water supply mains (flushing) leading to measurable amounts of iron, manganese, aluminium, PAH and turbidity (DWI *Annual Report*, Eastern Region, 2004). Anglian Water was fined £3,750 and ordered to pay costs of £11,106 as well as required to submit plans for the mitigation of damage caused and prevention of similar lapses in future.

One of the major themes of this volume is the argument that whilst markets surely have a place in the management of our increasingly scarce water resources, the neoliberal ideology of "privatise, privatise, privatise" is facile and dangerous. Private markets cannot properly regulate a resource which is, properly thought, a universal heritage, that is to say a *common*. Privatising common resources tends to result in a few people or corporations making large profits whilst taxpayers are forced to pony up to make such market interventions attractive. Currently one of the best examples of water privatisation involves the explosive growth in the bottled water market (Royte, 2008, and Chapter 6). As Maude Barlow points out:

> …those areas of life thought to be common heritage of humanity for the benefit
> of the many, now coming under corporate control for the benefit of the few
> (rich).

Even highly privatised water sectors like England and Wales require very careful and expensive regulation – a situation only likely to be compounded by what I call (in Chapter 6) the "securitisation" of water. What is needed is not more private market involvement based on the now-stale ideology of the free marketeers, but rather redoubled efforts to establish a system of water governance at global, regional and national scales. These issues are taken up in Chapters 2, 3 and 6.

Water: A Confluence of Challenges

As sketched out in this opening chapter it is clear that European water managers in the 21st century face a complex variety of interacting challenges. Though Europe is broadly speaking well supplied with adequate water resources to meet its needs, there are regions of chronic shortage as well as regions of unacceptably low water quality. Where either of these are the case there are strong arguments for a pan-European approach to balance out surpluses and deficits, especially among EU current and prospective members. Indeed, the EU's role as a water manager at the transnational scale has grown and developed considerably in the 1990s, originally based on the need specified in the *Treaty of Rome* to foster the economic development of member states. And it is difficult to development economically without adequate water resources. Subsequently, as we shall see in subsequent chapters, the EU has articulated a much more proactive and comprehensive water management policy based on a bringing together of its numerous specific Directives on things like bathing water quality and nitrate pollution.

The EU also has a role in managing specific instances of transnational water management, as ongoing programmes for the Rhine, Elbe and Danube Rivers clearly shows. Though Europe does not suffer from the sorts of dangerous "hydropolitics" that characterise water use conflicts in other parts of the world, complex and difficult decisions do need to be made. For example, can upstream riparian states be convinced to agree to curtail their own industrial uses of water

> *"Water is the basis of life, and our stewardship of it will determine not only the quality but the staying power of human societies."*
> Postel, 1992

and to invest in expensive tertiary wastewater treatment in order that downstream states can have access to a sufficiently abundant and clean water resource?[13] More subtle perhaps are the environmental and health impacts of our current use of water. While there are strict standards at EU and national levels for most of the known pollutants in our water (The Drinking Water Directive specifies maximum allowable concentrations, "MACs", for over 60 known pollutants), relatively little is actually known about the interaction effects of the multiple pollutants that have turned some water courses into chemical stews. Even the alteration of characteristics as subtle as the velocity and temperature of water can have significant effects on human health if they alter the ways in which particular chemicals are held in aqueous solutions.

Clearly there are also issues to be addressed related to the differential usage of resources. Above we saw that citizens of different countries use vastly different amounts of water, much of it inefficiently. This is particularly so in the countries of the development, especially the US and Canada. One simple action that accounts for nearly 1/3 of water use is toilet flushing – switching to low flow toilets would instantly save a great deal of water without necessarily incurring any loss of

---

13 "Riparian" refers to countries with frontage along watercourses. "Riparianism" is an established principle of water allocation law. See Chapter 2.

function or convenience. See the table at left for more data about how water use was structured in the UK in the 1990s.

In terms of industrial usage, the story is one of progressively reduced usage per unit of production over the last century or so. Currently, the largest industrial users of water include thermal power production (coal-fired plants especially), metallurgy (especially aluminium production) pulp and paper production, and municipal sanitation. For example, roughly 15% of industrial water usage is in metallurgy, 4% in pulp and paper and 5% in automobile production. Put another way, each litre of beer requires about 5 litres of water, each motor car roughly 5,000 litres, each tonne of steel 8,200 litres, each tonne of paper 500 litres.

Environmental impacts are another area where the EU has sought to take the lead in establishing a common European programme for water management. Adopting agreed standards for things like flood control are especially critical on the continent where one country's flood defence scheme can actually cause problems for countries downstream. More complex will be dealing with water management problems that are caused by EU or national programmes in other areas. Environmental audits of the Common Agricultural Policy for example have repeatedly shown that it encourages profligate water use in irrigation and the pollution of those waters with chemical fertilisers and pesticides. Thus the EU must now attempt to reconcile one set of flagship initiatives intended to promote greater efficiency and economic productivity in agriculture with another set of flagship initiatives intended to improve the quality of the environment. River Basins such as the Guadalquivir in Andalucia, Spain are tragic case studies of the dire consequences for land and water quality attendant on CAP-induced shifts to extensive cereals production in the 1980s and 1990s.

Finally, the EU and its member states have a broader responsibility to the global community to manage their own water resources more prudently. In particular member states must now begin the difficult process of rethinking the complex relations between water and standards of living and development, for it is no longer acceptable for us to shift the negative externalities of our profligate use of water in the domestic, industrial and agricultural spheres onto less developed countries. As noted above this deplorable situation arises out of many complex and poorly understood processes including:

- The insatiable Western appetite for exotic and out of season foods which results in the export of "virtual water" from less developed countries to more developed countries;
- The pernicious export of inappropriate Western water technologies to less developed countries, usually under the duplicitous guise of "aid";
- Our palliative rather than solutions-orientated approach to water and health issues in less developed countries. It is surely unconscionable that while we can be motivated to provide rapid generous aid in the aftermath of an outbreak of cholera or typhus or of a disaster like flood, we lose interest and political will in addressing the underlying problems that exacerbate these conditions in the first place.

The charge to Europe's water managers therefore is not just a parochial or regional one; rather it is global in scope because the hydrological cycle is ultimately also global.

Suggested Activities for Further Learning:

This introductory chapter has necessarily been wide-ranging and complex and this may create the impression that there is too much to do, too much to think about. Yet if those of us involved in aspects of water management adopt the credo of "thinking globally, acting locally", it is much easier to see what we can do in our own professional spheres to address some of these important issues. In this section there are a number of suggested activities designed to help readers and/or students develop their understanding of the issues introduced in this chapter. Most take the form of "research, thinking and writing" projects, though a few involve a more precisely targeted analytic approach to specific technical problems. By contrast the Suggested Activities for some subsequent chapters may seem overly technical and "narrow".

Suggested Activities for Learning:

1. Write an essay about how global water balances have changed over the past 20 or 40 years. Have some areas, once water rich, become progressively water-impoverished? Or is the current situation one characterised by long term stability? Where can one access good quality statistics about global water resources, consumption and shortages? Is water shortage purely a matter of climatological and geographical "bad luck"?

2. Write about the history of water supply in two different European countries or regions. Are there important similarities or differences in the timing of water system development, their economic organisation, the role of the state, etc.? Are there long-term trends in water supply organisation that have implications for the short and medium term future of European water supply?

3. Examine one country's or region's attempts to implement Integrated Water Resource Management. What have been the successes and/or failings of these initiatives? Are there "lessons for the future" to be drawn from these case studies?

4. Prepare a short explanation of how the "qanat" system operated in the Middle East and/or the "acequia" system in Spain and Latin America. Particular attention could be paid to the institutional arrangements that allow such systems to function, rather than the physical infrastructure of canals, tunnels, reservoirs, etc.

5. Collect and collate a number of different estimations of the minimum amount of water needed to sustain human life each and every day. Consider the variation in estimations and argue for the one you consider most acceptable.

Chapter 2
Water Regulation at the Global Level

Introduction

The purpose of this chapter is to explore the idea, advanced by Varady et al. (2009), Karkainen (2004) and others that the "aperture of water governance" has widened beyond local, regional or even national initiatives. In the previous chapter I sketched out the very early history at administering water, but Varady and others are suggesting that recently there has been nothing short of a new "ontology of global water initiatives". Whether these constitute, or come to constitute, a system of hydro-governance rooted at the global scale remains to be seen. Yet, given the persistence of water shortage around the world and indeed its intensification by both climate change and the political economy of "virtual water" exports from already dry countries, there is clearly a prima facie need for a global response. I would go further, arguing there is an increasingly desperate need to colonise Varady's emergent "aperture" with a new "ontology" founded on ineluctable human and social rights to water.

In this chapter I start by examining a number of cases of inter-state contestation over water resources (though none of these cases are exemplars of so-called "water wars"). I then turn to the historical development of binational and multinational treaties and agreements designed to regulate competing uses of water (e.g. navigation on inland waterways), culminating in the 1992 Rio Summit on Sustainable Development. The Rio Summit, as we shall see, marked something of a pivotal point in the evolution of water governance for a number of reasons, including the articulation of broad principles (rather than specific treaty terms) and the recognition that non-state, non-governmental, organisations had a significant role to play. Yet as of the end of the first decade of the 21st century it remains an open question whether or not these advances mark the development of even the foundation for truly global water governance.

Looking at the management of water resources from the point of view of communities or even single countries runs the risk of implying that these levels of administration are somehow entirely autonomous with respect to water management decisions. But of course this is not so, if only for the simple reason that *water moves*. Unlike minerals or vegetation which are fixed in location throughout their lifecycle, water, by its very nature, moves: it moves from the mountains to the oceans, via a complex network of lakes, rivers, underground streams, and wetlands. Even in land-locked bodies such as lakes water moves – through the mechanisms of the *hydrological cycle* (evaporation and evapotranspiration) and, though the mechanisms are a little harder to understand at first (some of these processes are raised again in our examination of water quality management in Chapter 4), through temperature-related layer processes. All of this is undeniable scientific fact, and would perhaps even be mundane, were it not for the observation that *a great many watercourses cross national boundaries*. Sometimes more than once. When this happens there can be conflict over the allocation of water use rights, particularly where there is economic growth and development occurring on one or both sides of the boundary. Consequently the United Nations has declared that:

> Since rivers, lakes and aquifers – and the hydrological cycle as a whole – cross national frontiers, water policy planning must, to some extent at least, be carried out *within an international framework* (UN, 1979 cited in McDonald and Kay, 1988, p. 50, my italics).

Of course none of this suggests just *how* conflicts over (scarce) water resources can be ameliorated. As it happens reference is often made to the pattern or relationships that have existed in the past, thus tending to privilege historical (and geographical) inertia in contemporary hydropolitical relations. In water law recourse is made to two different principles for the management of international water relations and the adjudication of disputes: the doctrines of "prior use/appropriation" and "riparianism". The former suggests that the allocation of water rests upon the fundamental legal maxim "first in time, first in right." The first person to use water (called a "senior appropriator") acquires the right (called a "priority") to its future use as against later users (called "junior appropriators") simply by virtue of having been there first. "Riparian" rights are the rights of landowners to use water that is on or adjacent to their property. Landowners may have riparian rights only if their land touches some body of water, for example by having a riverside location. Landowners generally also have rights to groundwater underneath their property that are similar to surface water rights. Between them, these two legal principles define the two key ways in which water allocation disputes are articulated and adjudicated, and it follows that the principles themselves can clash. In both cases however the guiding principle may be obscured by the fact that state and inter-state institutions generally insert themselves between landowners and water resources by asserting a licensing authority.

Historically the development of international water management regimes have always been closely associated with the parallel development of interstate political systems. Thus watercourse issues were included in pretty much every major international treaty of the 19th century (Caponera, 2007). For example the Final Act of the Congress of Vienna in 1815, which formally ended the Napoleonic Wars, included several articles concerned with the management of international watercourses. In that Act these were defined as rivers or lakes "of concern" to two or more states – and who need not be riparian. This conception of hydropolitics has been periodically re-articulated up to the present day; at the Berlin Congress in 1885, at Versailles in 1918, and at those more recent treaty-making convocations discussed below. Indeed, it is possible to find hints of this modern conception of water governance as far back as the 16th century Treaty of Westphalia, which ended the Thirty Years' War and established a distinctly modern foundation for inter-state relations. There is not the space here to recount more fully the fascinating history of international water law, but it is worth noting that in general the codification of such has coalesced around the following principles (Caponera, 2007, p. 193):

- A duty to cooperate and to negotiate with a genuine intention of reaching an agreement;
- A prohibition of management practices likely to cause substantial and lasting injury to other states;
- A duty of prior consultation;
- An equitable utilisation of shared water resources.

These principles have subsequently formed the basis of legal "tests" in cases of contestation of water development plans especially at the International Court of Justice (ICJ) at The Hague. In the next chapter we shall look at one such case especially relevant to European water management; the Gabcikovo-Nagymaros Diversion scheme for the Danube River, decided by the ICJ in 1997. Here we will consider only a few different examples of interstate cooperation (and occasional contestation) or water resources insofar as they suggest both an emergent crisis and a global level of water governance.

Hydropolitics: Water-related Security Issues

So-called "hydropolitics" has emerged onto the international diplomatic scene as a key threat to the stability of the global political economic order (Ohlsson, 1995; Lowi, 1995). Water resources in certain regions of the globe are under increasing pressure (remember the statistics about water scarcity from the previous chapter), and this pressure is resulting in a new and dangerously belligerent attitude to the allocation of water resources, especially in watersheds that are shared amongst a number of states. And since most watersheds are shared, the scope for conflict over water is large indeed. Speaking about the countries of the Nile Basin in 1985, former Secretary General of the UN Boutros Boutros Ghali said: "the next war in our region will be over the waters of the Nile".[1] A large part of the problem is rooted in the tendency to identify sovereignty over "sufficient" water resources with the territorial integrity of the modern state itself. Thus the founder of modern Israel, David Ben-Gurion, declared,

---

1 He has restated this view as recently as February 2005 in a BBC interview with Lyse Doucet.

> it is a necessity that the water sources, upon which the land depends, should not
> be outside the borders of the future Jewish state,

since any other situation would create a source of political tension in the region. Subsequent to its foundation in 1947 water developments such as the National Water Carrier have been an integral element of Israeli state policy – and a source of friction with neighbouring states in particular Palestine, Syria and Jordan (de Chatel, 2007; Lowi, 1995). Of course not all disagreements over water result in armed conflict; indeed most are resolved through a combination of compromise, third party mediation, institution-building (remember our brief discussion about the work of Eleanor Ostrom on common property institution building in the previous chapter) and political strong arm tactics (e.g. Turkey's uncompromising position on the impacts of the Greater Anatolia Project on downstream states). Nevertheless there can be no doubt that hydropolitics will become more, not less, important in the 21st century, even in relatively peaceful and prosperous regions such as Europe and the UK.

The study of hydropolitics is a relatively new academic pursuit (Allan, 2001; el Zain, 2007; Ohlsson, 1995; Wolf, 1995). As scholars and researchers become involved in studies of water-related issues, they quickly realised that water is a multifaceted resource. As life is in fact impossible without water, the latter is increasingly acknowledged as a resource of inestimable value, especially for those who don't have it. What is even more pertinent is the fact that the human species has affected every possible ecological space on the planet, probably in an unsustainable fashion. This is the fundamental driving force behind hydropolitics as more people compete for and rely on water resources that are themselves being degraded and depleted. But these problems are complex and multidimensional and therefore require multidisciplinary approaches.

From a multidisciplinary perspective, it becomes evident that hydropolitics is about:

- conflict and cooperation;
- involving states as the main actors;
- shared international river basins; and
- the socio-political identity of water itself.

Yet, a glance at the literature mentioned above shows that such a state-centric focus on conflict and conflict mitigation in shared international basins is not the only focal point of hydropolitical interest. Meissner (1999, pp. 4-5) has also tried to define hydropolitics. He sees the study of hydropolitics as

> the systematic investigation of the interaction between states, non-state actors
> and a host of other participants, like individuals within and outside the state, and
> co-operation within the framework of the state.

There are some notable exceptions to this state-centric tendency. Swatuk and Vale (2000) challenge the notion of state centrism in a refreshing but not yet mainstream approach. Allan (2000, p. 233) criticises this prevailing International Relations perspective for failing to consider the dampening effect of the trade in water-rich products – what he calls "virtual water" (Allan, 1998) – that has been partly responsible for preventing once confidently predicted water wars from occurring (Bulloch and Darwish, 1993; Gruen, 1992; Starr, 1991). Moreover it is necessary to think carefully about the issue of *spatial scale* in hydropolitics, as not all water disputes are international, or even inter-state in character. For example, the ongoing conflict over the waters of the Colorado River Basin has both intra-state and inter-state dimensions, with seven US states as well as Mexico involved.

Another critical element in the understanding of hydropolitics is the *range of issues* that are covered. These can best be understood as a horizontal dimension of the discipline of hydropolitics. In reality, the range is infinitely wide, including issues such as conflict and its mitigation, states and non-state actors, water service delivery, water for food, the social value of water, the political value of water, the psychological value of water, water demand management (WDM), water as a target of aggression (e.g. the deliberate targeting of water infrastructure during the NATO bombardment of Yugoslavia in 1999), water as an instrument of peace, water and gender, water and ecosystems, and water as a critical element in sustainable development. To complete the analytical picture we can identify spatial scale as the vertical dimension of our enriched conception of hydropolitics.

In this section we will look briefly at three different case studies of "hydropolitics", though not of the more journalistically interesting kind involving armed conflict. The first case study is of what is perhaps one of the world's longest running set of institutions for the resolution of international water disagreements: the US/Canada Transboundary Water Agreement dating back to 1909. The second case study involves the long-running dispute between the US and Mexico over the waters of the Rio Grande and Colorado Rivers. The third case study examines the deep-rooted and ongoing dispute between riparian states along the Nile River and its tributaries in north-eastern Africa.

Case Study: The US/Canada Transboundary Waters and Columbia River Agreements

Most of North America is drained by three major river systems: the Fraser-Columbia in the Pacific Northwest, the Missouri-Mississippi in the central part of the continent, and the St Lawrence in the eastern zone. The hydrogeography is organised such that some of Canada's pristine lakes and rivers eventually become sources of drinking water and conveyors of sewage for Americans hundreds and perhaps thousands of miles to the south. Indeed, by the 1950s and 1960s proposals were being mooted for water development projects of a truly breathtaking scale, including the (in) famous North America Water and Power Alliance (Micklin, 1977; Reisner, 1993). Discussed by a subcommittee of the US Congress in the

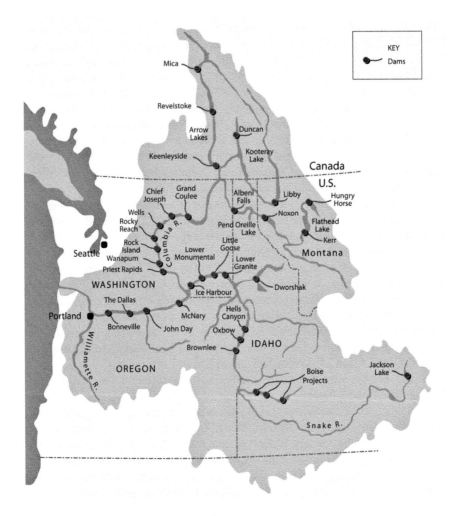

Figure 2.1   Columbia River Basin

Source: US Army Corps of Engineers, http://www.nwd-wc.usace.army.mil/report/colmap. htm.

mid-1960s the NAWAPA would eventually have affected more than 2 million square kilometers, a land area roughly twice the size of the province of British Columbia and eight times the size of the UK.

It would have involved several hundred separate dam, diversion and power plant projects, astronomical cost and would have been by far the largest water development project ever undertaken.[2]

---

2   US weekly magazine *Newsweek* hailed the NAWAPA plan as "the greatest, the most colossal, stupendous project in history" and numerous key water managers and politicians

Recognising the potential for inter-state conflict over water resources, the US and Canada agreed the *Boundary Waters Treaty* in 1909.[3] The key achievement of that treaty, which continues in force to this day (though much augmented) was the establishment of the International Joint Commission to investigate controversial issues. Though its recommendations are not binding it is widely respected by lawmakers on both sides of the international boundary and acts as a clearing house for information to do with boundary water issues. Since 1909 there have been more than two dozen water-related treaties between the two countries, stretching from coast to coast all of which are managed by the IJC. Although some criticise the IJC for its consensual approach and inability to enforce rulings it is noteworthy that fewer than 25% of its decisions have split along national lines and most of its recommendations find some level of implementation.

Based on the principles contained in the BWT, the parties were eventually able to negotiate the *Treaty Relating to Cooperative Development of the Water Resources of the Columbia River Basin* (*"Columbia River Treaty"*). The *Columbia River Treaty* explicitly recognised that the construction and operation of three projects in Canada would increase both the useable energy and dependable capacity of power plants in the US, as well as provide irrigation and flood control benefits in the US, all of which would not be possible at the same cost without the three BWT/CRT-condoned projects (Figure 2.1). In return for building the three *Columbia River Treaty* projects in Canada, the Treaty specifically entitled Canada to a lump sum payment for various downstream (flood control and power generation) benefits, as well as one half of the additional power generated by power plants in the US that resulted from water storage across the border in Canada (Paisley, 2002).

From the Canadian perspective, one of the reasons that this agreement was possible, and indeed logical, was because of the neocorporatist model of water management that has long obtained in the province of British Columbia. The BC Hydro Corporation, a publically-owned "crown" corporation, was established by a pro-business government in 1962 and BC water management has been largely within its exclusive purview. Thus it has been able to negotiate on behalf of British Columbian interests as a parastatal organisation (with due recognition of respective federal and provincial jurisdictions). As suggested by the name, the mandate of BC Hydro has been to develop and expand hydropower resources, which it has done, with alacrity. Recent research by Glavin (2002) shows how this quintessential hydro-corporation (in some ways like the Tennessee Valley Authority discussed elsewhere in this volume), has systematically diverted, channelled, impounded and over-abstracted the province's water bounty. In recent years it has even brokered so-called "Independent Power Producer Contracts" (IPPCs) that effectively privatise sections of BC rivers for private power producers.

---

both sides of the border expressed their support for the project. Fortunately, the public mood was very much against any such idea by the mid-1970s and the idea was shelved.

3   On Canada–US state practice in the field, see P. Wouters, "Theory and Practice in the Allocation of the Non-Navigational Uses of International Watercourses: Canada and the United States: A Case Study," *Canadian Yearbook of International Law*, 43 (1992).

Case Study: The Mexico–US Colorado River Water Dispute

The 1944 *United States–Mexico Treaty for Utilization of Waters of the Colorado and Tijuana Rivers and of the Rio Grande* allots to Mexico a guaranteed annual quantity of water from these sources. The treaty does not provide specifically for water quality, but this did not constitute a problem until the late 1960s. Rapid economic development and increased agricultural water use in the United States spurred degradation of water quality received by Mexico. The United States began diverting significant amounts of water from the Colorado River in order to irrigate new areas under cultivation. The Wellton-Mohawk Irrigation and Drainage District of Arizona was the most important of these projects. At the same time that excess water became scarce, Arizona began pumping highly saline drainage from the Wellton-Mohawk project back into the Colorado River. While the United States continued to fulfil the Treaty's water quantity requirement by returning most of the diverted water to the river before it reached Mexico, it chose to overlook the decline in water quality. Much of the greatly increased water abstraction went to irrigate export crops. With a view to resolving the problem, Mexico protested and entered into bilateral negotiations with the United States. In 1974, these negotiations resulted in an international agreement, interpreting the 1944 Treaty, which guaranteed Mexico water of the same quality as that being used in the United States.

The resultant international agreement between Mexico and the United States resulted in substantial cost to the United States in the amount (estimated) of US$280,000,000: for (1) desalting of the Wellton-Mohawk Drainage Project; (2) extension of a lined bypass drain to carry Wellton-Mohawk drainage and brine from the desalting plant to the Santa Clara slough on the Gulf of California; and (3) financing Mexico's improvements and rehabilitation of the Mexicali Valley.

Simultaneous with the US–Mexico dispute there has been a long-standing "internal" hydropolitical struggle over division of the waters of the Colorado River between the seven riparian US states strung along its length from its headwaters in Wyoming and Colorado to its lower reaches in southeastern California and southwestern Arizona, before it enters Mexico and empties into the Gulf of California. The historical origin of this problem can be traced back to a US Supreme Court ruling that the doctrine of "prior appropriation" applied to all surface waters west of the Continental Divide; a ruling that was immediately interpreted as a threat to future water rights by upstream states (Cech, 2005; Reisner, 1993). Moreover, state water managers were anxious that an inter-state solution should be negotiated rather than one emanating from the federal government. Thus an inter-state pact, dividing up the Colorado's waters, was negotiated and signed in late 1922. Unfortunately, two things became increasingly clear in subsequent decades. First, urban and economic development has continued to be highly asymmetric in the Colorado basin, with California and Nevada accounting for the lion's share of development, and therefore water demand. Second, it turns out that the 1922 pact was based on hydrological data pertaining to a relatively wet period, rather than longer term mean flows such that the Colorado was significantly over-allocated (by up to 70% according to some estimates). In the late 1990s, with Lake Powell and other reservoirs in danger of drying up the US federal government pushed for a new agreement based on better, more up to data hydrological data and on the need to conserve as well as allocate the Colorado's waters. The new pact, signed in late 2007, is innovative inasmuch as it includes specific incentives for states to undershoot their permitted abstraction rights (i.e. to conserve) and formalizes a de facto trade in water rights that has been going on since the 1920s.

Case Study: The Nile River Basin

From its major source at Lake Victoria in east-central Africa, the White Nile flows north through Uganda and crosses the border into Sudan. After a journey of several thousand kilometres it eventually meets the Blue Nile in the dusty heat of Khartoum (capital of Sudan) which, by that time, has made its precipitous descent from the Ethiopian highlands. From the confluence of the White and Blue Niles at Khartoum, the now "complete Nile River then flows 1,800 km northwards through the desert, into Egypt and on to the Mediterranean Sea.

Contained within the Nile catchment, encompassing something like 1/5 of the African continent and some of its most arid lands, is both a tremendous resource and a tremendous – and growing – demand. According to the World Bank, the Nile River Basin is home to an estimated 160 million people, while almost 300 million live in the ten countries that share the Nile's waters. Within the next 25 years, population within the Basin is expected to double, adding to the increased demand for water generated by growth in industry and agriculture. In the UN's World Water Development Report, Egypt is ranked 156th, Sudan 129th, Kenya 154th, Ethiopia 137th and Uganda 115th for per capita water availability. And, as Table 1.2 makes clear, the region's countries are almost all "water scarce" – that is, they must make do with considerably less than 1,000 m^3/capita/year. With population growth of approximately 3%/year tensions over increasing water scarcity have become a serious bone of contention among the 10 countries that share its basin – Burundi, Rwanda, the Democratic Republic of Congo (DRC), Tanzania, Kenya, Uganda, Ethiopia, Eritrea, Sudan and Egypt.

Trouble arises partly from two water allocation agreements signed during the (British) colonial era, the 1929 Nile Water Agreement and the 1959 Agreement for the Full Utilization of the Nile – These agreements gave Egypt and Sudan extensive and largely exclusive rights over the river's use. Upstream countries, including the East African countries of Kenya, Uganda and Tanzania have expressed concern over the long-standing arrangements, arguing that the treaties have served to give Egypt unfair control over the use of the river's waters (Hultin, 1995). Moreover it is also argued that the colonial treaties are illegitimate since they ignored the rights and needs of upstream countries. Upper Nile countries such as Zaire, Rwanda and Burundi were never party to the 1929 or 1959 treaties and therefore consider their riparian rights to be undiminished. Egypt and Sudan, on the other hand, have been reluctant to renegotiate the treaties and this has, at times, strained relations between the upper- and lower-riparian nations. In the 1970s and 1980s Egypt stated baldly that it would be willing to go to war to protect its exclusive access to Nile waters as enshrined in the 1959 treaty. Climate change, and the realisation that the Nile may not actually be able to sustainably deliver the 84 billion cubic metres specified in the 1959 treaty have further exacerbated matters.

During the 1990s, attempts to resolve mounting disagreement surrounding the Nile Basin and the allocation of its waters led to the development of a regional partnership approach. In 1998 all the riparian states except Eritrea began discussions with a view to creating a regional partnership to better manage the Nile Basin. A transitional mechanism, the Nile Basin Initiative (NBI), was officially launched in February 1999 in Dar-es-Salaam and in 2002 a permanent secretariat was established based in Uganda. Burundi, Sudan, Tanzania, Uganda, the Democratic Republic of Congo, Egypt, Ethiopia, Kenya and Rwanda are all involved in the NBI, with Eritrea participating as an observer.

According to Antoine Sendama, one of the Nile Basin Initiative's regional coordinators, the 10 countries which share the Nile and its sources work to find a way of cooperating on developing the Nile Basin sustainably and effectively. Most countries in the region share a similar history of poverty, high population growth, environmental degradation, unstable economies and insecurity. All are blighted by poverty and many by recurrent civil war including Sudan, Ethiopia, Eritrea, Uganda, Rwanda, Burundi and DRC. Within the NBI there are plans to harness the basin's waters for irrigation, and also the establishment of an energy policy to provide power for all the countries in the region. Indeed the regional approach embedded within the NBI will be key to its success – the vision statement for the NBI calls for "sustainable socio-economic development through equitable utilization of, and benefit from, the common Nile Basin water resources" (NBI, 2009). Currently the NBI is coordinating a growing portfolio of water development projects encompassing irrigation, hydropower generation, drinking water supply and sanitation, transport and flood alleviation.

However, critics of the NBI have argued that the initiative has been a closed affair in which only the states involved and the World Bank, which provides much of the funding, have had any influence, largely ignoring the voices of ordinary. Elizabeth Birabwa, a writer on environmental issues, has said that there is hardly any information flowing between the NBI secretariat and the media, because the language used by the secretariat was "too technical and distanced from us":

> Ugandan MPs have raised an issue that affects ordinary people. But the issues are shrouded in secrecy, big moneys being spent, some of it to be repaid by the people who live along the Nile, but the people know nothing (Science in Africa, 2002).

Civil society groups like IUCN and local NGOs have also criticised the running of the NBI and have formed a parallel initiative they say would enable them to participate more equally in the NBI process. As elsewhere in the world governments have been slow to welcome the inclusion of civil society into the NBI.

Hydropolitics Around the World

Elsewhere in the world, Israel, Syria and Jordan have a well-known and long-standing dispute over the waters of the Sea of Galilee and the Jordan River, Lebanon, Iraq and Syria have all complained about Turkey's plans for the Tigris and Euphrates Rivers and Pakistan, India and Bangladesh have a long-running conflict involving the major rivers of the subcontinent (Lowi, 1995: Ohlsson, 1995). One potentially bright spot in the "hydropolitical world" is the Mekong Basin Initiative, through which the countries of the Mekong River Basin in southeast Asia seem to be emerging with an innovative model for collaborative water management. Few of the world's major rivers are contained within a single country and this simple fact creates the need to develop negotiated allocation mechanisms that avoid the potential disasters of an appeal by one riparian state or another to violence (remember Hardin's "tragedy of the commons" idea from the last chapter). In a rather sober assessment of future prospects Ismail Seregeldin of the World Bank declared that "the wars of the 21st century will be fought over water" (de Villiers, 2000, p. 15).

Water management is, at the beginning of the 21st century, a highly complex and contested enterprise. Water has been subject to human regulation since before even Roman times, and is now regulated by a large number of different organisations, agreements, and processes operating at different spatial scales. Adding the complexity is the fact that not all agreements have the status of "law"; "declarations", "compacts", "treaties", and "agreements" are not cut from the same legal cloth and can often further cloud the issues. In the following sections I will introduce readers to this sometimes bewildering network of regulation, and also discuss the growing role of non-governmental organisations (NGOs) in within it.

International Environmental Law and Treaties

The last generation has seen the emergence of ecological degradation as an issue of global concern, and environmental lawmaking has been the primary mechanism used to promote natural resources conservation, pollution control, and other forms of environmental protection (Wouters, 1997). International Environmental Law (IEL) has grown to encompass hundreds of international and regional treaties, many thousands of national laws, disseminated by more than 180 countries and multilateral organisations such as the UN agencies, international development banks and environmental non-governmental organisations (NGOs). Every new environmental treaty is preceded by a multitude of conferences, backroom negotiations, position papers drafted by government officials and affected interest groups, intensive lobbying by many governments, multilateral organisations, environmental NGOs, scientists, trade groups and industry representatives, and anyone else who has a stake in the outcome. Every ratified IEL treaty is then followed by periodic conferences of the parties, reports by the managing secretariat(s), intergovernmental panels and advisory committee meetings, and the same kinds of active lobbying efforts that characterise international environmental lawmaking processes. By any measure of diplomatic and legal activity, the field of IEL has experienced remarkable growth and become highly prominent since the 1972 Stockholm Declaration on the Human Environment.

According to Patricia Wouters, a world expert on international water law, the genesis of modern international water law can be traced back to the post-war period. In the early 1950s, international bodies, such as the Institute of International Law (known by its French acronym IDI)[4] and the International Law Association (ILA),[5] began studying the law applicable to water disputes and arguing for its extension and deepening. Each of these has made important contributions to the development of law relating to international waterways. Their early work culminated in the Helsinki Rules (1966), supplemented, *inter alia*, by a number of subsequent resolutions, such as the Montreal Rules on Marine Pollution (1982), the Seoul Complementary Rules (1986) and ultimately the UN Watercourses Convention (1997). The overarching general principle of the ILA's work on international water law is contained in Article IV of the Helsinki Rules which provides that the principle of "equitable utilization" governs the use of the waters of international drainage basins – one of four global water governance principles noted at the outset of this chapter.[6] The Helsinki Rules have played an important role in the codification and progressive development of this branch of international law. According to Paisley (2002), the principle of equitable utilisation requires states to act reasonably and equitably when dealing with transboundary water resources in their territory. It requires that the reasonableness of any utilisation is to be determined by weighing all relevant factors and by comparing the benefit that would follow from the utilisation with the injury it might inflict on the interests

---

4 The Institute of International Law, more commonly referred to as l'Institut de droit international (hence IDI), founded in 1873, is comprised of a restricted number of international lawyers who produce reports and resolutions on various topics of private and public international law.

5 The International Law Association has a broader membership than the IDI, but also meets regularly and produces reports and adopts resolutions aimed at codifying and progressively developing various topics of public and private international law.

6 Article IV of the Helsinki Rules provides: "Each basin State is entitled, within its territory, to a reasonable and equitable share in the beneficial uses of the waters of an international drainage basin" – this is therefore an IEL expression of the "riparian" doctrine, albeit one that raises the contentious issue of "equitability".

of another basin state. The genius of the principle of equitable utilisation lies in its flexibility because it prescribes a "reasonableness" test for determining what is lawful or unlawful conduct in connection with international water resources. Note that this international legal principle already takes us away from the rote application of earlier principles such as "prior use" (first in time, first in right) or "strict riparianism".

Returning briefly to earlier comments made in Chapter 1 on the idea of a "global commons", remember what I said about how they all provide some kind of *common* resource, be it the fish in the sea, the minerals of the earth, the air we breath or the water in the oceans and freshwater courses. The overexploitation of these resources has long been recognised as something that needs to be addressed and as such there have been many international conventions to try to organise the use of these resources through international environmental law. Of course there are serious questions about the effectiveness of international treaties and conventions to discipline the behaviour of even well-meaning signatories. Despite the plethora of treaties, conventions, NGOs and other institutions that have arisen largely since the first Earth Day in 1970, many global regions are increasingly wracked with what are now called "hydropolitical" or "hydrosocial" problems (Ohlsson, 1995; Swyngedouw, 2005). In some regions (e.g. the Jordan River Basin) things may soon reach a point where water becomes an explicit and immediate security issue, raising the prospect of Security Council and/or NATO interventions. We do not accept Lomborg's (1995) facile claim that because there have not yet been any armed conflicts *solely* caused by water scarcity there will not be such in future.[7] Moreover, in dealing with the realities of contestation over water resources we are pushed to consider carefully the emergence of a meaningful and effective global level of water governance, constituted out of international treaties, state or quasi-state organisations and non-governmental groups.

In the remainder of this chapter we will look at just a few of these treaties: what they deal with; who has not ratified them as opposed to who has; and if they are actually doing anything to alleviate the environmental problems we are imposing on the planet. The scope of notes on individual pieces of treaty-making vary widely, mostly because the author has tended to pick and choose amongst the panoply of agreements for the purposes of making targeted analytical points rather than being comprehensive. Readers in need of a comprehensive overview are directed to Wouters (1997), Paisley (2002) and Caponera (1992).

To date something like 150 international treaties concerning water have been signed, with the vast majority being *bilateral*, that is to say treaties between only two (riparian) nations. Only two dozen truly *multilateral* treaties have been ratified and entered into force (Caponera, 2007; Conca, 2006). Overall the historical

---

7 Of course as a neoliberal economist, Lomborg also rejects the idea that there ever will be a water shortage that cannot be solved by raising the water price to the "equilibrium" level. In the 21st century and in some parts of our world that may prove to be a very high price indeed!

trend over the past century has been towards the extension of laws and treaties originally designed to cover shipping and the building of major infrastructure (such as canals and dams) to all areas of water development and to make them universally binding (Dellapenna, 2001). Agencies involved in this process have included the United Nations and its legal research directorate, the International Law Commission, multilateral institutions such as the European Union and third party non-governmental organisations (NGOs) such as the International Law Association.[8] The most recent achievement of real significance is the 2004 "Berlin Rules on Water Resources" which attempts to provide a codification of emerging international law relating to surface and ground waters (ILA, 2004). As we shall see in the next chapter, European water management law has closely followed these international trendlines, particularly with respect to the massive and very complex Water Framework Directive passed by the Commission and Parliaments in 2000.

Prior to the 20th century transnational water regulation was primarily a bilateral affair, although even as far back as the post Napoleonic War settlements there was attention to protecting the right of non-riparians to navigate along waterways. By the early 20th century however there was a growing sense of internationalism generally, and specifically (for water) a realisation that institutions and legal structures needed to evolve at the global level to stave off potential conflict based on the clash of the competing doctrines of territorial sovereignty, riparianism and prior appropriation. As noted above, one of the earliest pro-active and prospective water agreements was the 1909 *Boundary Waters Agreement* between the United States and Canada. With over 8,000 kilometres of shared border, the St Lawrence River, the Great Lakes, the Red River of the North, the Columbia, Stikine and Yukon Rivers, there is much potential for conflict. The genius of the IJC established by the Agreement is that it is organised around the need to advise and reach consensus rather than imposing solutions – a power which would instantly have ramped up the politicisation of water issues between the two countries. The first major concrete accomplishment of the IJC process was the Columbia River Treaty of 1961 which, although 30 years in the making, allowed for the mutually agreed and beneficial development of the Columbia River Basin.[9]

Elsewhere in the world moves were made in the Post-World War I settlements to establish a transnational level of water governance. The first manifestation of this was the *Convention and Statute on the Regime of Navigable Waterways of International Concern* signed at Barcelona in April 1921. This international agreement revisited and clarified the rights to international navigation on international watercourses which had last been discussed at the treaty-making

---

8   The world's most important international legal NGO, founded in 1873.

9   Whatever one thinks of the environmental consequences of the Columbia River developments, it is remarkable that the process was able to support the peaceful development of such a significant resource – see Chapter 6 for environmentally orientated critiques of large water projects such as those on the Columbia River.

Table 2.1   Selected international water agreements

Title	Date	Purpose
Convention and Statute on the Regime of Navigable Waterways of International Concern.	1921	Discourages unilateral modifications to river basins, establishes internationalism as a key principle.
Helsinki Rules on the Uses of the Waters of International Rivers.	1966	Developed by the ILA as a standard for adjudicating water conflict, eventually transformed into the 1997 Framework Convention. Primarily important for the establishment of the principle of "equitable utilization".
Resolution 3129 (XXVIII) on Cooperation in the Field of the Environment Concerning Natural Resources Shared by Two of More States.	1973	Broadens and deepens the water-specific concerns of the 1966 Helsinki Rules.
Mar del Plata Conference and Action Plan.	1977	Established that national governments should development national strategies for extension and improvement of water services.
International Drinking Water and Sanitation Decade.	1981-1990	Coming out of the UN Conference at Mar del Plata in 1977, established a decadal water development drive that brought water services to over 1 billion people. 2005-2015 has been designated as another water development decade.
International Conference on Water and the Environment.	1992	Gives further impetus to the privatisation agenda by declaring that "water has an economic value in all its competing uses" (Principle 4).
The UN Conference on Environment and Development.	1992	Institutionalised the modern movement towards "sustainable development", including sustainable water management.
UN Convention on the Law of the Non-navigational Uses of International Watercourses.	1997	A major statement of international law on the sharing of international watercourses.
Second World Water Forum and Ministerial Conference.	2000	Over 5,000 delegates and 130 countries represented, articulated a wise use doctrine that ties water provision to further privatisation.
General Comment No. 15 (2002), The Right to Water, Substantive Issues Arising in the Implementation of the International Covenant on Economic, Social and Cultural Rights.	2002	Re-affirms universal right to water.
Berlin Rules on the Use of Water.	2004	Developed by the ILA as a state of the art, comprehensive codification of customary international water law covering both surface and ground waters.

Source: Adapted from the United Nations database on international treaties.

conventions which ended the First World War only a few years before. In a nutshell the Barcelona Convention guaranteed the rights of *third parties* as well as riparians to navigate international waterways and in particular to limit the power of riparians to obstruct or otherwise unduly regulate traffic on "watercourses of international concern" (Caponera, 2007, p. 213). Whilst riparians retained the right to impose national systems of watercourse regulation, in the context of the recent European war, it made sense to reaffirm a fundamental right of all to reasonable access and utilisation. This is of course in line with the principle of customary law recited above, that of the right to "equitable utilisation of water resources", subsequently expanded and amplified (Paisley, 2002).

The next key milestone in global water governance was the agreement of the Helsinki Rules on the Uses of the Waters of International Rivers, signed in Helsinki in 1966. This has become *the* touchstone document for the codification of the "equitable utilization of water resources" and "no harm" principles. The Helsinki Rules also adopted an *inclusive* interpretation of territorial sovereignty which in effect imposes obligations on all states to ensure that other water users are not unduly disadvantaged by water utilisations that affect either quantity or quality of water available (Caponera, 2007). The UN Framework Convention of 1997 and the Berlin Rules of 2004 substantially update and strengthen the Helsinki Rules, but most of the literature on international water law marks them as one of the pivotal moments in the development of 20th century water governance.

Convened at the resort of Mar del Plata, Argentina in 1997, the objective of the Mar del Plata Conference was,

> to promote a level of preparedness, nationally and internationally, which would help the world to avoid a water crisis of global dimensions by the end of the present century.

Specifically it was intended that the conference would establish a principle of the human right to water subsequently enshrined in a number of related conventions and treaties but never unambiguously articulated in an enforceable form. The expectations of Mar del Plata were as follows:

> It is hoped that the Water Conference would mark the beginning of a new era in the history of water development in the world and that it would engender a new spirit of dedication to the betterment of all peoples; a new sense of awareness of the urgency and importance of water problems; a new climate for better appreciation of these problems; higher levels of flow of funds through the channels of international assistance to the course of development; and, in general, a firmer commitment on the parts of all concerned to establish a real breakthrough so that our planet will be a better place to live in (Mageed, 1978).

The Conference approved an action plan, which was officially called the Mar del Plata Action Plan. It was in two parts: recommendations that covered all the essential components of water management (assessment, use and efficiency; environment, health and pollution control; policy, planning and management; natural hazards; public information, education, training and research; and regional and international cooperation); and 12 resolutions on a wide range of specific subject areas. The key achievement of Mar del Plata was to explicitly articulate for the first time a universal human right to water; something which has subsequently been the subject of much controversy (Biswas et al., 1997; Gleick, 1999):

> …all peoples, whatever their stage of development and their social and economic conditions, have the right to have access to drinking water in quantities and of a quality equal to their basic needs (United Nations, 1977).

As Peter Gleick (1999) points out, while the right to water had been implicit in a range of previous international agreements such as the 1948 *Universal Declaration of Human Rights*, it was only at Mar del Plata and after that such a right was expressed explicitly, albeit with little attention to enforcement.

One of the resolutions of the Mar del Plata Conference was to designate the 1980s as the "Decade for Clean Water" with the bold goal of bringing water to everyone however poor.[10] Whilst more than 1 billion people were reached over the course of the decade, global inequality grew even faster. By the end of the decade more than a billion people were still without access to clean water – a situation ironically similar to the beginning of the decade. Choguill, Franceys and Cotton (1993) comment that:

> Despite the failure to meet the quantitative goals, much was learnt from the experience of the water and sanitation decade. There was further realisation of the importance of comprehensive and balanced country-specific approaches to the water and sanitation problem. Most importantly, perhaps, was the realisation that the achievement of this goal that was set at the beginning of the decade would take far more time and cost far more money than was originally thought.

The decade 2005-2015 has now been designated as the "second clean water decade" with the specific objective of meeting the water-related Millennium Development Goals agreed in 2000. By 2008 the interim "report card" for water services presenting a mixed picture: though great strides have been made, it is still true that more than a billion people remain without access to clean water and sanitation facilities (United Nations, 2008).

---

10   There is of course considerable debate about what quantity constitutes a reasonable minimum standard of provision – Gleick (1999) discusses the range of choices between 25 and 50 litres/person/day whilst WHO adopts the latter for development and emergency aid applications.

The 1992 Rio Earth Summit was heralded as the turning point for global environmental policy. More than one hundred countries came to the Rio summit, which sought to merge two critical international concerns – environmental protection and economic development – that had been evolving on different tracks during the 1970s and 1980s. For developing countries, the merger of environment and development was a major improvement over earlier environmental conferences and provided hope for increased North-South cooperation. In addition, the Cold War had recently ended, and the rise of a one superpower world meant that East-West conflicts would not dominate this conference, as they had earlier international environmental efforts. On paper, at least, the Earth Summit did provide a potential vision for moving toward sustainable development – that is, toward both greater environmental protection and greater economic justice. The Earth Summit yielded two legally binding treaties: the Framework Convention on Climate Change; and the Convention on Biological Diversity. Also a product of the Summit were a set of nonbinding general principles known as the "Rio Declaration", a set of nonbinding principles covering everything imaginable, including water management, and the blueprint for sustainable development entitled Agenda 21. The assembled governments also established the Commission on Sustainable Development (CSD) to integrate environment and development into the UN system while providing a forum to monitor the implementation of summit commitments. Agenda 21 encourages investment in water treatment as a more sustainable option, especially in the developing world where up to 80% of all diseases and a third of deaths are caused by drinking contaminated water:

> Water is needed in all aspects of life. The general objective is to make certain that adequate supplies of water of good quality are maintained for the entire population of this planet, while preserving the hydrological, biological and chemical functions of ecosystems, adapting human activities within the capacity limits of nature and combating vectors of water-related diseases. The multi-sectoral nature of water resources development in the context of socio-economic development must be recognized, as well as the multi-interest utilization of water resources for water supply and sanitation, agriculture, industry, urban development, hydropower generation, inland fisheries, transportation, recreation, low and flat lands management and other activities (Agenda 21, Chapter 18).

The key problem with Agenda 21 of course, is that it is a *process*, rather than a set of strictly binding agreements, about rethinking the relations between economic development and environmental management. As political philosopher Tim Luke (1998) notes, the idea of "sustainable development" does to at all abandon the modernist idea of unlimited development, but in fact revalorises it, just so long as a little more attention is paid to the "sustainability" of economic growth. Its most lasting achievement, to date, has been in terms of the language it has introduced and the impetus it has given to ideas about inter-generational equity and the rights of non-human actors to a healthy and sustaining environment.

Just before the Rio Summit, five hundred participants, including government-designated experts from a hundred countries and representatives of 80 international, intergovernmental and non-governmental organisations attended the International Conference on Water and the Environment (ICWE) in Dublin, Ireland, in January 1992. At its closing session, the Conference adopted this Dublin Statement and the Conference Report. The problems highlighted are not speculative in nature; nor are they likely to affect us only in the distant future. They are here and they affect humanity now. The future survival of many millions of people demands immediate and effective action.

The Conference participants called for fundamental new approaches to the assessment, development and management of fresh water resources, which can only be brought about through political commitment and involvement from the highest levels of government to the smallest communities. Commitment will need to be backed by substantial and immediate investments, public awareness campaigns, legislative and institutional changes, technology development, and capacity building programmes. Underlying all these must be a greater recognition of the interdependence of all peoples, and of their place in the natural world.

Principle 1: "Fresh water is a finite and vulnerable resource, essential to sustain life, development and the environment".

Principle 2: "Water development and management should be based on a participatory approach, involving users, planners and policy-makers at all levels".

Principle 3: "Women play a central part in the provision, management and safeguarding of water".

Principle 4: "Water has an economic value in all its competing uses and should be recognized as an economic good".

> (Guiding Principles. The Dublin Statement
> on Water and Sustainable Development)

Unfortunately it is Principle 4 which has garnered most attention, implying as it does the necessity of further privatising the world's water resources (this issue is taken up in more detail in Chapter 6).

Whilst the Helsinki Rules have been very influential, even a legal lodestone, they have never been formally constituted as binding global rules for water management. Consequently there has been a concerted drive, led by the ILC, to further codify, standardise and clarify global water law. In 1997 the ILC submitted the UN Convention on the Law of the Non-navigational Uses of International Watercourses to the UN General Assembly for approval. As Wouters (1997) points out, "at the centre of the discussions were issues relating to the identification of the substantive rules that determined states' rights and obligations over watercourses:

- What limits apply to watercourse States' entitlements to use transboundary waters?
- How are the various factors to be weighed in the overall assessment of an equitable and reasonable use?
- Where a conflict of uses arises, what rule(s) determines which use should prevail?
- What role does "harm" play and how is it to be assessed?"

As always the response of member states to these questions varied in accordance with their location relative to key watercourses: downstream and non-riparian states tended to take environmental protection and equal utilisation positions, whilst upstream riparian states re-affirmed their commitment to riparianism and first use. Structurally these issues haave of course been seen before: indeed the three case studies presented earlier in this chapter show how the creation of credible and neutral institutions can broker deals where all parties win and lose to some extent. The final text of the Framework Convention on the Law of the Non-Navigational Uses of International Watercourses was adopted by the UN General Assembly on 21 May 1997 but has yet to come into force. Article 5 of the Convention declares:

> Watercourse States shall in their respective territories utilize an international watercourse in an equitable and reasonable manner. In particular, an international watercourse shall be used and developed by watercourse States with a view to attaining optimal and sustainable utilization thereof and benefits therefrom, taking into account the interests of the watercourse States concerned, consistent with adequate protection of the watercourse.

This article clearly attempts to balance both riparian and non-riparian states' rights and obligation as well as the needs of the "silent partner" to the Convention, the environment itself. Interestingly the Convention does not establish a separate implementation Secretariat (as was done with the US–Canada Boundary Waters Treaty, the Nile Basin Initiative and many other water-environmental agreements and treaties) but tries to establish a set of generic rules for all UN Member states. Irresolvable disputes can be taken to the International Court of Justice at The Hague, as was the Gabcikovo-Nagymaros dispute (though not under the Framework Convention) discussed in the next chapter.

In 2002 the United Nations issued General Comment No. 15 (2002), The Right to Water, Substantive Issues Arising in the Implementation of the International Covenant on Economic, Social and Cultural Rights. This is not a treaty per se, but comments on and clarifies the 1976 International Covenant on Economic, Social and Cultural Rights (ICESCR), reaffirming that:

> the human right to water entitles everyone to sufficient, safe, acceptable, physically accessible and affordable water for personal and domestic uses. An

adequate amount of safe water is necessary to prevent death from dehydration, reduce the risk of water-related disease and provide for consumption, cooking, personal and domestic hygienic requirements (Paragraph 2).

The very wide-ranging ICESCR had been ratified by 149 countries by June 2004 and is therefore legally binding upon all UN member countries (unlike the 1997 UN Convention of Water). This official "general comment" on the ICESCR means that the right of each and every person on earth to an amount of water adequate to support their personal development at an affordable price is legally enforceable in effect reinforcing the intention of the 1977 Mar del Plata treaty.

The International Law Association, which had previously worked to develop the 1966 Helsinki Rules, developed the 2004 Berlin Rules not as a new or separate set of rules for managing transboundary water issues, but as a state of the art summation of almost a century of modern law-making and extension of the Helsinki Principles to all waters above and below ground (and not just transboundary waters). Unlike the 1997 Framework Convention which does seek to become binding, the Berlin Rules operate more like a practitioner's guide than a substantive new statement and it will be interesting to see how the two may interact, especially given the failure to date to achieve sufficient national ratifications to allow the 1997 Framework Convention to come into force. Significantly the Berlin Rules strongly restate the principle, emanating from natural justice, of a universal right to water (this time without any explicit mention of price) alongside suggested mechanisms for enforcement.

As you can see, the world of international water regulation is a complex one indeed, and we really have only scratched the surface of it! As this is not a text on environmental law, we must leave the details of this legislation to the experts (cf. Caponera; Conca; McCaffrey; Paisley; Wouters). However, we can make a few key observations here. International water law by the 21st century concerns the rights and obligations that exist, primarily between States, for the management of transboundary water resources. Such legal rules and principles are dedicated to preventing conflict and promoting cooperation of shared water resources. The by now well-established principles of "equitable utilisation" and "no harm" are now being expanded and extended into principle of sharing downstream benefits of water development (Paisley, 2002). Declarations of vague principles covering global environmental issues do nothing to actually provide sufficient water for drinking and sanitation and other crucial needs of human beings in the foreseeable future. States need clear and enforceable guidelines now. That this is a difficult and complex task is evidenced in the length of time it has taken the United Nations to deal with the matter and the fact that very many countries remain divided over key issues – as Wouters (1997) points out it is just about possible to infer from the global hydrological map just how a given country will vote on global water issues.

Of course this is a highly selective listing of international agreements making direct reference to the assertion or acknowledgement of "rights" over fresh water resources. As with so many international agreements and conventions however there is seldom any clear and unambiguous enforcement mechanism contained in these texts and consequently they can be seen merely as laudable statements of principle. Enforcement currently seems to depend as much on moral suasion by specific countries or by the rising tide of environmental NGOs which seek to promote water development around the world. It is to a brief consideration of these organizations that we now turn.

Non-governmental Organisations (NGOs) and Water

In addition to the complex web of international agreements, treaties, etc which bind states into an interlocking system of mutual responsibilities discussed in the previous section, global water regulation has also been greatly affected by the emergence, in the last thirty years, especially of an influential and well-developed network of non-governmental organisations (NGOs) devoted to monitoring environmental conditions and improving environmental management. Among the most famous of these are Greenpeace and Friends of the Earth, but there are also a number of important technical/professional NGOs specifically orientated towards water issues and water management. In this section we will look at the emergence of the environmental NGO sector in the last thirty years in the context of increasing globalisation and restructuring of the nation-state and then introduce a number of the most important water NGOs currently operating. As in the preceding section, the emphasis here is on an indicative rather than exhaustive treatment of international water NGOs, that is to say those non-governmental, non-state organizations which are increasingly inserting themselves into the rapidly evolving system of global water governance.

Environmental NGOs: A Brief History

The term "NGO" was coined by the Economic and Social Research Council of the United Nations in 1946, which needed a specialist term for the increasing number of non-state organisations which it had increasingly to deal. While perhaps increasing in number at the time, such organisations certainly pre-date the invention of the term (e.g. the Royal Society for the Protection of Birds was originally founded in the 1890s), and the term did not come into general use until much more recently. The UN maintains a watching brief on the expansion of the NGO sector, at least at the global level, and can be consulted by people interested in learning more about which NGOs are working in specific countries or issues areas (e.g. water). Some examples, largely taken from the list of 1,600 or so organisations that have some sort of status with the UN, include: the International Federation of Red Cross and Red Crescent Societies, CARE International, Catholic Relief Services, Worldwide Fund for Nature (WWF), Royal Society for the Protection of Birds (RSPB), Confederation of Independent Trade Unions in Bulgaria, AARP (American Association of Retired Persons), Amnesty International, American

Bar Association, League of Women Voters, Girl Scouts, Rotary International, Romanian Independent Society for Human Rights, etc., etc. I should note the 1,600 figure is by no means comprehensive as even small states such as Bulgaria may have in excess of 200 NGOs devoted to the environment, apart from those devoted to social, political, religious, cultural or other issues. Therefore, any reckoning of the number of environmental NGOs around the world would probably rise into the tens of thousands – though of course the vast majority would likely be small and very local in orbit and influence (Cellarius and Staddon, 2003).

In the social sciences there is a general understanding that the growth of the NGO sector since the late 1960s coincided with the decline of the welfare state model of government (Wolch, 1989). Thus, as successive governments (especially in the west) have withdrawn from numerous areas of social, political, cultural and environmental life, NGOs have often formed to fill the vacuum left. While part of this process can be seen to have been largely benign in other areas the situation has been more problematic. In the environmental sector, for example, the withdrawal of the state from many direct front-line regulatory functions has coincided with a Reaganite/Thatcherite affirmation of the power of the "free market" to properly regulate environmental "goods" and the simultaneous denigration of "society". Thus, water regulation has been drastically reorganised in the 1980s and 1990s in the largest shake-up of the vertically-integrated public sector water structures since their establishment in the 19th century. As the state has withdrawn from the water sector, and other sectors, over the last twenty years then, there has been a concomitant rise in the number of "quango", "public-private partnership" NGO or other 'para-statal' arrangements. In many ways, what we have witnessed is the emergence of a "shadow state" which, viewed optimistically, "could also signal a very different configuration of the political economic system more in line with the classical pluralist vision" (Wolch, 1989, p.218).

As hinted above, the term "NGO" refers to a large and diverse body of private organisations that exist in the space between government agencies and the for-profit commercial sector (Cellarius and Staddon, 2003). It is a somewhat fuzzy term that defines organisations largely in terms of what they are not. Here are a couple of definitions from important international development organisations, starting with the UN (since it invented the term) after all. According to the United Nations:

> A non-governmental organization (NGO) is any non-profit, voluntary citizens' group which is organized on a local, national or international level. Task-oriented and driven by people with a common interest, NGOs perform a variety of services and humanitarian functions, bring citizens' concerns to Governments, monitor policies and encourage political participation at the community level. They provide analysis and expertise, serve as early warning mechanisms and help monitor and implement international agreements.

The World Bank defines NGOs as:

private organizations that pursue activities to relieve suffering, promote the interests of the poor, protect the environment, provide basic social services, or undertake community development.

Also, the World Bank usually speaks of non-governmental organisations (NGOs) by which it means non-profit organisations (NPOs)and community-based organisations (CBOs) that are

(i)   entirely or largely independent of government;
(ii) not operated for profit; and
(iii) exist to serve humanitarian, social or cultural interests, either of their memberships or of society as a whole.

In wider usage, the term NGO can be applied to any non-profit organisation which is independent from government. NGOs are typically value-based organisations which depend, in whole or in part, on charitable donations and voluntary service. Although the NGO sector has become increasingly professionalised over the last two decades, principles of altruism and voluntarism remain key defining characteristics.

At the global level the NGO sector is quite diverse, with the differentiation of purpose, organisational structure and relations to client groups helping to create a messy sort of "Alphabet Soup" (Cellarius, 2004). Thus business orientated NGOs are know as BINGOs, big organised NGOs as BONGOs, government run NGOs as GRINGOS (or QUANGOS), etc. While the typology introduced above is somewhat comical, there are a few specific variables which allow the interested outsider to begin to understand the internal structure and differentiation of the sector:

1. *Scale*
   – International, intermediary (i.e. between international organisations or other donor and grassroots), local (but be careful about what local means, who is being represented);
   – North vs. South or East vs. West (postsocialist).

2. *Purpose*
   – Charity, relief, development, political;
   – Member support (e.g. AARP, cat lovers' society to import cat food without incurring customs duties imposed upon businesses) vs. public interest (e.g. Public Citizen working on consumer safety legislation and regulations);
   – Single vs. multiple interest.

3. *Main Activities*
   - Research, lobbying, programme management, education, technical assistance;
   - Networking, umbrella groups that link numerous NGOs, often on the basis of some shared characteristic;
   - Operational vs. advocacy.

4. *Inspiration or Affiliation*
   - Church (Latin America), government, business, Greens.

This is not the place for a detailed analysis of the linkages between the NGO sector and political ideas about "civil society" (cf. Cellarius and Staddon, 2002; 2003), but surely it is ironic that, just as the post-Thatcherite consensus proclaims the need for communities to sort themselves out, their capacity to do so is so craftily circumscribed. Indeed, it is arguably the case that official efforts to support civil society, primarily through legal and fiscal means, but also as we shall see discursively, may contribute to the foreclosure of grassroots democratisation and civil society development through the generation of exclusive cadres of ENGO professionals, intermediary organisations, and the official representatives of recognised social problems (Sampson, 1996; Simonson, 2003). Canadian activist and author Maude Barlow (2007) has decried this double movement which, she suggests, is the key threat to the emergence of a truly democratic "world water covenant". With better access to outside resources and support, the professional or "official" water NGOs (like the Global Water Partnership) increase their likelihood of survival, while the grassroots groups struggle. Thus the analysis developed in this paper holds the important policy implication that governments ought not to so hastily normalise in public policy what is in fact a highly ideological and hegemonising model. Rather government ought to ensure that there is room for the autonomous expression of local political will and energy through inclusionary rather than exclusionary policies.

Water NGOs: A Short List of the Most Prominent or Well-known

Table 2.2 contains a listing of some of the most prominent specifically water-orientated NGOs. It is immediately apparent that most of these organisations have been around for less than 20 years and in fact reflect the emergence of water as a global issue. Of course these organisations are differently powerful. Some, such as the International Water Association and the World Water Council have quickly become significant players on the international stage, often by sponsoring industry-side events. Others such as the International Rivers Network and Lakenet are prominent because of their work publicising poor water management practices, dam projects, etc. around the world.

It can be readily seen that these organisations can be divided into three groups: those organised around campaigning against poor water management

Table 2.2   Selected international water NGOs

Organisation	Brief Mission Statement
GWP – Global Water Partnership	The Global Water Partnership (GWP), established in 1996, is an international network open to all organisations involved in water resources management: developed and developing country government institutions, agencies of the United Nations, bi- and multilateral development banks, professional associations, research institutions, non-governmental organisations, and the private sector. GWP was created specifically to promote Integrated Water Resources Management (IWRM) which aims to ensure the coordinated development and management of water, land, and related resources by maximising economic and social welfare without compromising the sustainability of vital environmental systems.
International Water Association	IWA is in a better position than any other organisation in the world to help water professionals create innovative, pragmatic and sustainable solutions to challenging global water needs. IWA is at the forefront in connecting the broad community of water professionals around the globe – integrating the leading edge of professional thought on research and practice, regulators and the regulated, across national boundaries and across the drinking water, wastewater and storm water disciplines.
International Water Resources Association	Since its official formation in 1972, IWRA has actively promoted the sustainable management of water resources around the globe. IWRA is committed to the sound management of water resources through advancing water resources and related environmental research, promoting water resources education, improving exchanges of information and expertise, networking with other organisations who share common interests and goals and providing an international forum on water resource issues.
Water Aid	WaterAid is the UK's only major charity dedicated exclusively to the provision of safe domestic water, sanitation and hygiene education to the world's poorest people. Its funding comes from a precept levied on all water companies in England and Wales and on private donations.
WaterWatch	Set up in February 1994, WaterWatch is a voluntary network of campaigners, groups and individuals, concerned with all water issues (consumer, environmental, regulatory and health) in England and Wales.
World Water Council	The mission of the WWC, formed in 1994, is to raise public awareness about the critical water issues at all levels including the highest decision-making level organise the triennial World Water Forum.
International Rivers Network	Established in 1985 the International Rivers Network protects rivers and defends the rights of communities that depend on them. IRN opposes destructive dams and the development model they advance, and encourages better ways of meeting people's needs for water, energy and protection from damaging floods.
LakeNet	LakeNet is a global network of more than 900 people and organisations in 90+ countries working for the conservation and sustainable management of lakes. The LakeNet Secretariat is a US-based nonprofit organisation dedicated to bringing together people and solutions to protect and restore the health of the world's lakes.
International Water Law Project	Created and directed by Gabriel Eckstein, the mission of the International Water Law Project (IWLP) is to serve as the premier resource on the Internet for international water law and policy issues. Its purpose is to educate and provide relevant resources to the public and to facilitate cooperation over the world's fresh water resources. As the subject evolves and develops, the IWLP will continue to update its pages and databases.
The International Network on Participatory Irrigation Management (INPIM)	The INPIM is a global network for the promotion of people, public, private partnerships in irrigation and water resources management. The network has over a thousand members actively involved in the irrigation sector and four national chapters in Albania, India, Pakistan and Indonesia.

(IRN, Lakenet) or for a particular kind of good water management (GWP) and those attempting to capture a professional community (IWA, IWRA) and those attempting to capture global water discourse through dominating its central events or institutions (WWC). They are also quite diverse in their organisational structures, with only a few (such as Lakenet and IRN) being true "membership" organisations with any clear accountability towards that membership or constituency. Most others are highly professional "subscription" organisations, such as IWA, IWRA and WWC. All are united however in attempting to encourage certain forms of action at the global level, as well as in water "hotspots" around the world. Thus, IRN was especially active around organising opposition to the World Bank-sponsored Sardar Sarovar dam projects in India in the 1990s (Cullett, 2007; Shiva, 2002) and Water UK has worked especially to provide access to clean drinking water in poor villages in Africa South America and Asia. WWC, on the other hand, was established in 1996 as a vehicle for organising and communicating the needs of the burgeoning private water sector.

Particular note should be taken of the World Water Forum, which in little more than a decade has emerged as a powerful clearinghouse and focal point of global water development issues. The World Water Forum was instituted in 1996 as a triennial global water conference primarily orientated towards the burgeoning private water sector around the world. The March 2000 WWF in The Hague was the first time that a ministerial conference was held in combination with the otherwise private sector World Water Forum and the practice has been followed at the Third (Japan, 2003) and Fourth (Mexico, 2006) WWFs. Over 5,000 participants attended the event and there were no shortage of communiqués issuing from the Forum itself and from delegations. More important than any statements of novelty (which there were not), is the fact that the WWF is fast becoming a key site of water policymaking both intra and inter nationally. Indications are at the next event, in Istanbul in 2009 will be even bigger and more "powerful" – a kind of Davos for water.

Concluding Thoughts

This has been a necessarily wide-ranging chapter. The task of introducing the main contours of global water regulation has required us to move from an initial diagnosis of the need for a transnational or global level of water governance to the more obvious institutional dimensions of international environmental and water law, through to the environmental NGO sector, which we saw is becoming much more important in global environmental, including water, regulation. Whether or not these developments themselves constitute, or augur the emergence of, water governance at the global scale is debateable. On the one hand there has been no straightforward up-scaling of water governance from the "tried and true" system of regulation at the scale of the nation-state (whether internally or as a contracting party in hydro-treaty making). Thus, there is nothing easily recognisable as "global water governance", although Varady et al. (2003) would certainly caution that the search for similar homologies is misdirected and that a sensitivity to emerging networks of governances is needed:

> Ultimately, determining how best to achieve sustainability in multilevel water governance starts with a fundamental shift in the ontology of water governance: seeing GWIs as situated and operating within vital networks (Varady et al., 2003, p. 155).

Such a perspective suggests (to social scientists at any rate) a fascinating corollary: that the upscaling of concern to regulate key drivers of contention with respect to water resources could be occurring simultaneously with a reconstitution of the nature of governance itself. This then is the battle that Maude Barlow enjoins us to engage – the struggle over the regulatory identity of water itself: as a private good to be doled out to those who can pay, or a common property to which each and every inhabitant of the planet should have an enforceable right.

Suggested Activities for Further Learning:

Once again this has been a wide-ranging chapter, though necessarily so. Possibly the most useful "Activities for Learning" focus on the legal and institutional dimensions of global water management and on specific case studies of good/bad or successful/unsuccessful water management at the international scale:

1. Examine one or other of the international treaties or conventions having immediate bearing on the idea of a "right to water". How well is this right established? Are there enforcement mechanisms or precedent case law? What could be done to make the right to water stronger?
2. Study the structure, history and current activities of any of the large global water NGOs.
3. Discuss the contention that until regional water conflicts "go military" there is little that the existing and complex structure of global water regulation can do.
4. To what extent can the developments discussed in this chapter be said to constitute an incipient global level of water governance?
5. Earlier in this chapter it was suggested that the Mekong Basin Initiative was a "potentially bright spot" in the world of hydropolitics. Investigate the MBI and sketch out its strengths and weaknesses.

Chapter 3
Managing Water in Europe

Introduction

In this chapter we examine contemporary European water management systems, including institutions, legislation, and policy. At the heart of the chapter is a comparison of the emerging EU-level water management system (since 2000 unified within the *Water Framework Directive*) with those of certain member states, in particular the UK. As we shall see, water management law and institutions have evolved a "Russian Doll" structure, such that national institutions are framed within EC Directives, which are themselves shaped by the emerging global level of water governance. Much of this chapter is devoted to the straightforward presentation of details about relevant laws, Directives, agencies and other institutions and "stakeholders" in water management, but it is hoped that these trends can be seen as part of an historical-geographical process that has its roots in the distant past. Indeed, although the attempt to create a pan-European water management framework is new, many of the stakeholders and even the key principles are not new at all. For example, the existence of privatised water companies is not a novel invention of the Thatcherite 1980s or even the laissez-faire 19th century, but goes back rather further into the misty past. Also, the idea that national interests are closely tied to control over water is also not new. Through a brief examination of water management prior to the post-war era we shall see that royal "patents" to private water companies were being granted as far back as the 17th century and commercial agreements concerning water supply go back at least as far as the 14th or 15th centuries. Regulation of Europe's waters has therefore developed over many centuries in response to both growing and urbanising populations and changing ideas about the optimal mix between private and public sector involvement.

A Brief History of European Water Management

Systematic water management in European countries is really a product of 19th century campaigns by social and political activists to clean up Europe's burgeoning cities. Social campaigners such as Sidney and Beatrice Webb complained to London city authorities that the lack of clean potable water supplies on the one hand, and the pollution of streets, drains and watercourses with rubbish and sewage on the other, had combined to make the cities of Victorian England actively dangerous to human health. Somewhat earlier there is the famous story of London physician Dr John Snow, taking matters into his own hands during the cholera epidemics of the 1850s.

Faced with official indifference to his suggestion that most cholera sufferers had taken water from the same (polluted) well in London's Soho, he had the handle of the pump in question cut off (Ward, 1997). New cholera cases in the immediate area dropped to nothing within days. Virtually from that moment onward there was an increasingly relentless drive to modernise urban water supply and sewerage systems, leading to the formation of a new wave of limited liability private water supply companies[1] and public water control authorities in the second half of the 19th century. These two anecdotes suggest that modern European water management was borne of the conjunction of social concern over access to clean water in burgeoning cities on the one hand and an emergent science of water and public hygiene on the other.

But formal arrangements for water supply in European cities certainly goes back much further than the 19th century. Leaving aside the elaborate arrangements often made for manorial estates and monastic buildings, and the systems (including the famous aqueducts) established by Roman administrators up to the end of the 4th century AD, it is nevertheless true that as early as the 13th century urban

Photo 3.1   Pont du Gard, near Nimes, France

Source: Author.

---

1   And indeed the very idea of a "limited liability" company has special relevance to the water sector, where, until recently, the use of lead in pipework rendered much "clean" public supply toxic – the 'limited liability company' is a legal mechanism for protecting water company investors from liability for such risks.

authorities were attempting the systematic organisation of public water supply. By 1236 the City of London had a lead piped water system that brought water in from the Tyburn (where today's Oxford Street is). In the 14th century Bristol's civic leaders assumed responsibility for the pipes that brought water from Ashley Hill down to the centre of settlement at the quayside, completing the pipe from Knowle to St Mary Redcliffe church which had been installed by Robert de Berkeley in the 12th century. At about the same time monastic orders in Southampton, Gloucester and other cities agreed to allow urban residents to access their own water supply networks and these were shortly thereafter purchased outright by city authorities and the networks extended. In the late 16th century Sir Francis Drake was charged (once he had sorted out a small matter involving the Spanish Armada!) with constructing Plymouth Leat to bring water directly into the centre of the city of Plymouth. By this time ingenious technical advances in waterwheel and water screw designs had made the moving of larger volumes of water easier and more efficient. Plymouth Leat itself had no less than six water-powered mills along its 17-mile length, bringing water from the River Meavy directly into the city centre for free distribution to citizens. As water supply systems became more sought after it became more common for companies of "merchant venturers" to be granted exclusive patents to develop reservoir and pipeline systems to supply growing English cities with water from further afield, such as Sir Hugh Middleton's New River Company, granted a "Royal Patent" (essentially a monopoly) in 1610 to dig a 38-mile "new" river to bring clean water from Hertfordshire into central London.

There are many more such examples of late 16th and early 17th century attempts to bring a sufficient *quantity* of water into England's growing cities, but what of water *quality*? After all while it was certainly an achievement to be able to provide water straight into the heart of Restoration or Georgian London, what can be said of its overall quality and the arrangements in place to secure it? Unfortunately not too much, as the long-standing practice of dumping refuse including domestic rubbish (including dead animals) and excrement (animal and human) into the streets to be flushed into the nearest watercourse continued to be practiced until well into the 18th century. Indeed, as the satirist Tobias Smollett observed in *The Expedition of Humphry Clinker*, published in 1771:

> If I would drink water, I must quaff the mawkish contents of an open aqueduct exposed to all manner of defilement from the Thames...human excrement is the least offensive part of the concrete, which is composed of all the drugs, minerals, poisons used in mechanics and manufacture, enriched with the putrefying carcases of beasts and men...this is the agreeable potation extolled by Londoners as the finest water in the universe.

Interestingly, some of the first government legislation enacted by the English government actually concerned itself with water quality, including a 1388 Ordinance against "nuisances which cause corruption" of water and air and a 1535

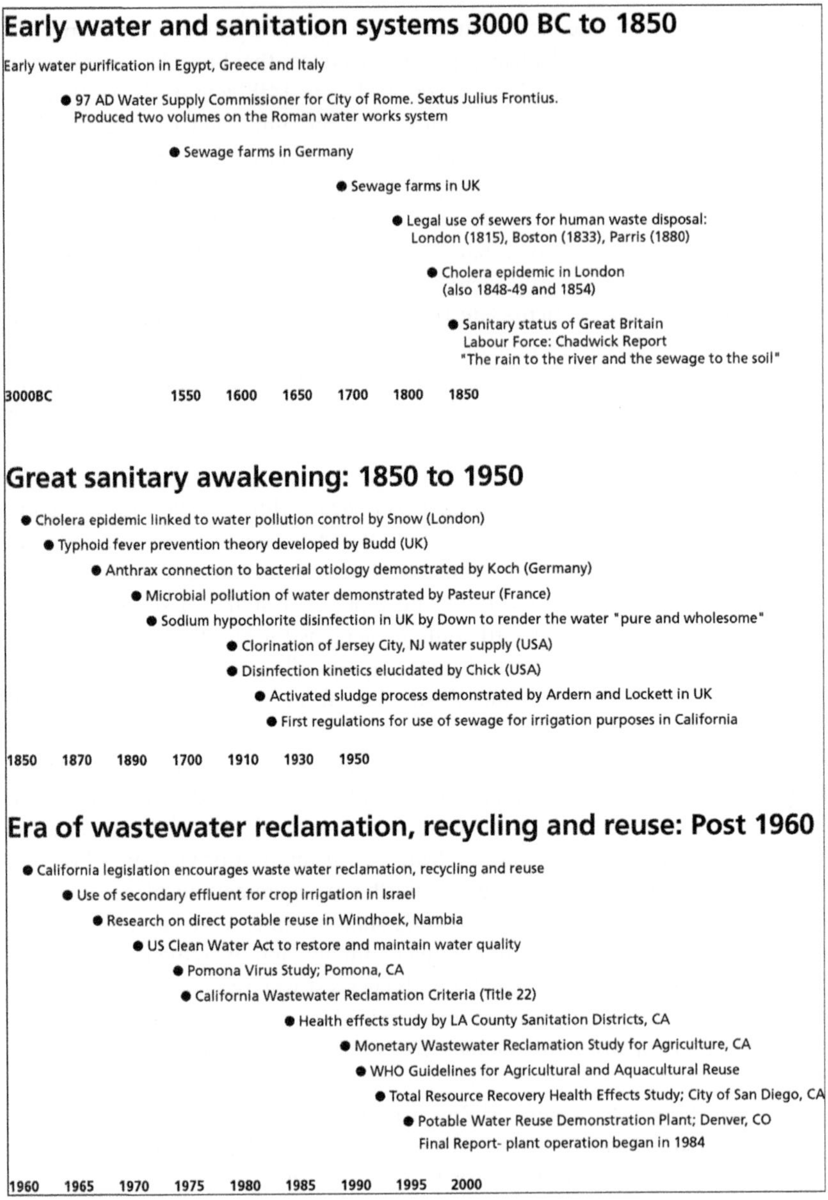

Early water and sanitation systems 3000 BC to 1850

Early water purification in Egypt, Greece and Italy

● 97 AD Water Supply Commissioner for City of Rome. Sextus Julius Frontius.
Produced two volumes on the Roman water works system

● Sewage farms in Germany

● Sewage farms in UK

● Legal use of sewers for human waste disposal:
London (1815), Boston (1833), Parris (1880)

● Cholera epidemic in London
(also 1848-49 and 1854)

● Sanitary status of Great Britain
Labour Force: Chadwick Report
"The rain to the river and the sewage to the soil"

| 3000BC | 1550 | 1600 | 1650 | 1700 | 1800 | 1850 |

Great sanitary awakening: 1850 to 1950

● Cholera epidemic linked to water pollution control by Snow (London)

● Typhoid fever prevention theory developed by Budd (UK)

● Anthrax connection to bacterial otiology demonstrated by Koch (Germany)

● Microbial pollution of water demonstrated by Pasteur (France)

● Sodium hypochlorite disinfection in UK by Down to render the water "pure and wholesome"

● Clorination of Jersey City, NJ water supply (USA)

● Disinfection kinetics elucidated by Chick (USA)

● Activated sludge process demonstrated by Ardern and Lockett in UK

● First regulations for use of sewage for irrigation purposes in California

| 1850 | 1870 | 1890 | 1700 | 1910 | 1930 | 1950 |

Era of wastewater reclamation, recycling and reuse: Post 1960

● California legislation encourages waste water reclamation, recycling and reuse

● Use of secondary effluent for crop irrigation in Israel

● Research on direct potable reuse in Windhoek, Nambia

● US Clean Water Act to restore and maintain water quality

● Pomona Virus Study; Pomona, CA

● California Wastewater Reclamation Criteria (Title 22)

● Health effects study by LA County Sanitation Districts, CA

● Monetary Wastewater Reclamation Study for Agriculture, CA

● WHO Guidelines for Agricultural and Aquacultural Reuse

● Total Resource Recovery Health Effects Study; City of San Diego, CA

● Potable Water Reuse Demonstration Plant; Denver, CO
Final Report- plant operation began in 1984

| 1960 | 1965 | 1970 | 1975 | 1980 | 1985 | 1990 | 1995 | 2000 |

Figure 3.1　Timeline of water management

Source: Author.

Act against the "annoying of the River Thames". There is little evidence however that such legislation had much real impact until both the population pressure on the resource and the science and technology of water quality management had developed considerably further. From the late 18th century onwards however advances in scientific understanding of water chemistry and pollution processes and hydrology led to new and more effective filtration technologies. In 1791 James Peacock patented a three tank settlement system which marked a significant improvement on earlier purification methods and presaged the arrival of mechanical filtration systems in the 19th century.[2] Sand filtration and other sorts of more sophisticated systems were developed in the first half of the 19th century by engineers such as Robert Thom and James Simpson, indeed Simpson's "slow sand filtration system" which depends upon the aerobic digestion of sewage wastes by diatoms and green algae growing on the sand surface is still in use today in the UK. Combined with sewage separation technologies such as the reinvented water closet such purification systems helped to solve one of the key water management problems of the 18th and 19th century – not how to get clean water to the city, but how to properly deal with the dirty water produced by city-dwellers!

The cholera, typhus and endemic diarrhoea epidemics of the 1830s and 1840s provided the next great push for improved public sanitation. With citizens, especially poorer ones, dying in their tens of thousands – more than 200 people are said to have died in Sunderland alone in 1831 – social reformers such as the Webbs, Edwin Chadwick and Sir Joseph Bazalgette began the process of transforming England's ancient and ill-adapted water sanitation systems into something truly modern. At this time the first comprehensive legislation to monitor and manage public infrastructure, the 1848 *Public Health Act* was passed, although its powers would not become other than voluntary until revision in 1875 – after this time local authorities had a formal duty to provide (minimal) water supply and sewerage.

Recurrent cholera epidemics caused by contact with sewage culminated in the "Great Stink" of 1858 when it became apparent to all that something drastic (and expensive) would have to be done about London's sewers. The very next year civil engineer Joseph Bazalgette convinced the government and the London Metropolitan Board of Sewers to begin the expensive process of constructing 83 miles of large "interceptor sewers" and over 1,000 miles of under-street sewers that would collect together all London's wastewaters and transport them to a distant estuary for effective removal from the urban zone (Halliday, 2001). This infrastructure, though now requiring much updating and improvement, remains the foundation of London's wastewater management system. Its construction required impressive feats of practical engineering, including the design of new vitrified bricks and underground building techniques that rank right up there with the construction of the railroads a generation earlier.

---

2 In fact the basic principle underlying Peacock's system, the idea of "settling" pollutants out by slowly moving contaminated water through settling tanks or ponds continues to be used today.

Photo 3.2   "Death's Dispensary" 1866 woodcut illustration depicting London's often deadly water supply

Elsewhere Europe's rapidly growing metropolitan areas were creating the same pressures for new solutions to the problems of potable water supply and sewerage removal. In Paris a system similar to London's was begun somewhat earlier, during the reign of Louis Bonaparte in the 1850s. This new system was the product of the genius of Baron Haussman, the same engineer who designed the system of boulevards that defines modern Paris. Victor Hugo even has his beleaguered hero Jean Valjean comment, in almost postmodern mien: "...Paris has another Paris under herself; a Paris of sewers; which has its streets, its crossings, its squares, its blind alleys, its arteries, and its circulation, which is slime, minus the human form." This is almost postmodern in its appreciation of the creation of a hidden circulatory network underground, underneath the more obvious above ground built environment (cf. Kaika, 2005). In Berlin, engineers took this hidden circulatory system further dividing the metropolitan area into "radial zones" each with its own system of sewage collectors feeding into ever-larger subterranean sewers. Berlin's system also required developments in pumping technology since the surface topography of the city does not vary enough for gravity to move wastewaters reliably.

Most importantly however, by the mid-19th century urban and national government authorities were quickly realising that it was very much in their interests to take a *systematic* view of the twin problems of potable water provision and wastewater removal. Scientific and technological advances also allowed for the realistic and effective collection and even treatment of ever-larger volumes of wastewater. Pastcur's experiments in France showed that there were many practical ways of purifying drinking water, and the development of the activated sludge process in Edwardian Manchester (by Ardern and Lockett) helped to make modern integrated water services provision a reality (Hassan, 1998).

Legal and Administrative Tools for Water Quality Protection

As discussed in Chapter 2 the legal and administrative means for protection of water resources comprise all sorts of regulations (bans, decisions, restrictions, treaties, etc.) in respect of activities which may affect water quality. A good many of these have been put in place at the international level, though as previously noted there are often significant problems with the enforcement of globally-mandated water standards. Much more effective are standards set at regional or national scales, even if these merely "adopt" international standards. Among this group of technical means the basic instrument is in the form of quantitative standards defining the admissible pollution concentrations that may be discharged into water depending on the type of recipient water (running water, stagnant water, sea water, sensitive/insensitive to eutrophication) and the source of pollution (municipal, agricultural, industrial, diffuse, point-source, etc.). Specific figures in respect of specific pollutants as well as their number and type of pollution parameters defined in those standards vary for different countries. US *Environmental Protection Act* (US EPA) limits and European Union Directives are examples of a standards-based approach to water quality management. Two such EU Directives are listed below to give you a sense of their organisation and logic:

- *Directive Concerning Urban Waste Water Treatment (91/271/EEC)*: Laying down the admissible values of pollutant concentrations (BOD_5, COD_{Cr} and TSS; for water classified as sensitive to eutrophication additionally total nitrogen and total phosphorus) discharged from municipal WWTP, requirements on the municipal wastewater treatment method (depending on the equivalent number of town inhabitants), recommended methods of defining the pollution parameters and the monitoring of compliance with regulations on effluent quality;
- *Directives on Emissions of Chosen Dangerous Substances Discharged into Water (76/464/EEC, 82/176/EEC, 83/513/EEC, 84/156/EEC, 84/491/ EEC, 86/280/EEC)*: Laying down the concentrations of "dangerous" substances (however defined) in discharged industrial wastewater. These directives also introduced admissible indices for pollutant emissions in relation to installed unit production capacity or unit of processed or used substance, thus encouraging improvements in production technology. Additionally these directives defined the admissible concentrations of these substances in surface water which forces consideration for the quality of effluent recipients (an ecological approach?) – it may happen that despite the compliance with quality standards for discharges, the standards for admissible concentrations in the recipient are exceeded; in such cases it is necessary to further reduce the volume of discharged pollutants in consideration for the recipient quality.

These and other EU Directives are, once promulgated by the European Commission, adopted into the national legal systems of member countries effectively "domesticating" them. This makes enforcement a national obligation (subject to the doctrines of proportionality and subsidiarity), but also helps to ensure that Europe's often (largely) shared waters are managed prudently *and consistently* in the interests of all.

In the UK the *water abstraction* and *water discharge* permits are the key related legal and administrative measures. These define the conditions to be met by discharged wastewater (quality parameters and the discharge regime) individually for every water user (source of pollution) as well as the conditions/ limits on abstraction from natural watercourses. Actual values for certain pollution indices defined in the water discharge permit may be elevated when the recipient has polluted excessively and the Environment Agency (in the UK, see below) will pick this up, either through regular monitoring or complaint. A common source of point source pollution in agriculture is careless management of sheep dipping, which can result in the release into watercourses of insecticides or other agents. In other cases polluters can be investigated and fined for discharged waters to natural watercourses that are outside standards agreed in the discharge permits.

Apart from such standards on pollutant emissions into water, the legal and administrative means for protection of water quality in European countries may comprise:

- Establishment of protection zones for water and water intakes (the general regulations on water quality protection are usually insufficient to ensure adequate protection of water intakes, particularly those for public water supply) and definition of limitations, bans and decisions in respect of the use of protection zone site and water use, e.g. ban on wastewater discharge and its agricultural use, ban on construction of residential estates, public roads, location of industrial plants, fuel stores, municipal and industrial waste dumping sites, animal breeding farms, cemeteries, ban on location of new water intakes and decisions on introduction of specific crop cultivation. The protective zone is usually bipartite, with the respective bans and decisions on specific uses being much more stringent in the internal part. In certain cases, e.g. for mountain streams and upper stretches of lowland rivers protective zones can cover entire river basin areas. The way of setting out the boundaries of protective zones is regulated by law.
- Restrictions on agricultural use of effluents and sludge. Such standards are focused on reduction or prevention of harmful impact of effluents and sludge on soil, vegetation, water, animals and people. *EU Directive (86/278/EEC)* is an example of such regulations; it determines i.a. the admissible concentrations of heavy metals in sludge, admissible amounts of these metals in the sludge-treated soil, and the admissible amounts of these metals which may be introduced into the soil in one year.
- Recommendations on farming focused on the reduction of non-point agricultural pollution. *EU Directive (91/676/EEC, The Nitrates Directive)* is a good example here: it identifies actions for reduction of water pollution by nitrogen compounds from agriculture. The directive recommends implementation of good agricultural practice (proper fertiliser use scheduling, suitable fertiliser dosage on slopes, at high groundwater levels, near water courses, retaining minimum cover of plants which draw nitrogen from the soil, preventing water pollution by surface runoffs and seepage beyond the root zone) and respective education.

There are also EC-mandated restrictions on marketing products, such as phosphate-containing detergents, that may pose significant hazard to the environment, including water. Technologies are now in place for both replacing the phosphates in their domestic and industrial applications and also for removing them from wastewaters, including urine-separation toilets (much phosphate pollution is related to human and animal urine), and biological removal processes (Hanæus, Hellström, Johansson, 1997). There is also a ban on the marketing and use of detergents (e.g. Directives 73/404/EEC, 82/442/EEC) if the average level of sensitivity to biodegradation of surfactants utilised in these detergents is lower than 90%.

Community policy, and environmental policy in particular, aims to achieve sustainability, transparency and fairness in water allocation and quality management. The principles underlying that policy are set out in Article 130r of the Maastricht Treaty and, for the sake of clarity, they are set out below, together with a brief discussion of each principle and how it applies to water policy. In the context of water management, sustainability requires that the level of protection of human health, of water resources and of natural ecosystems should be ambitious, aiming at a high level of protection rather than set at the minimum acceptable level – after all we simply do not know what the safe levels of exposure are to a large number of modern industrial pollutants over the long term. In accordance with existing EC undertakings, any water resource management system should also adhere to the following decision-making principles:

1. The Precautionary Principle: Any regulatory authority should, in the substance of its decisions, take into account the full range of knowledge and, where that knowledge is uncertain or incomplete, it should err on the side of assuming the worst possible consequence – *ergo*, it should avoid risks rather than maximise benefits. This principle recognises the moral duty to prevent damage to human health, the environment and to try to avoid doing so through ignorance or imprudence. It also recognises the difficulty and cost of reversing or rectifying damage to the environment. For example, once a sensitive aquatic ecosystem is compromised, in certain cases, it may be impossible to restore it. Once an aquifer is contaminated with pesticide residues, it may take decades to cleanse itself and, in the meantime, it may be unsuitable for use as a source of drinking water unless expensive treatment facilities are installed.

2. The Subsidiarity Principle: No regulatory authority should deal with an issue or make decisions about a policy that could be handled more effectively or more legitimately at a lower level of aggregation, i.e. at the level of Member States or their sub-national units. Conversely, no authority should occupy itself with an issue that cannot be resolved and implemented at a higher level.

3. The Principle of Transparency: No regulatory authority should take up an issue or draft a *projet de loi* that has not been previously announced and made publicly available to potentially interested parties not participating directly in its deliberations.

4. The Principle of Proportional Externalities: No regulatory authority should take a decision whose effects in financial cost, social status or political influence (especially for those not participating in it) are disproportionate either to the expectations inherent in their original charter or general standards of fairness in society. Long-term benefits and long term environmental consequences of nonaction must be fully taken into account, as must the precautionary principle. This has been the case in recent legislation such as the Urban Waste Water Treatment Directive and the Nitrates Directive

(both sketched out further below). It is further acknowledged in the use of "framework directives" which allow local solutions to local problems and often allow a higher ratio of benefits to costs.

As we shall see, these principals have been at least partially implemented with respect to the division of responsibilities for water management between the EU and its member states. However, participatory/democratic environmental governance is no panacea. Unless regulatory authorities are "properly designed", there is no reason to be confident that their decisions will be more sustainable or accepted as legitimate by those who have not participated in them and joined in the consensus which they are intended to promote. And, as emphasised above, governance arrangements never work alone but only in conjunction with community norms, state authority and market competition.

Water policy is an area which illustrates the need to have a coherent and effective coordination of all relevant Community policies. Where it is necessary for the protection of human health or where particularly dangerous or persistent pollutants are concerned, it is clear that common Community standards must apply. However, Community water policy must be sufficiently flexible to avoid the imposition of inappropriate or unnecessarily strict requirements simply for the sake of "harmonisation". Such flexibility would also ensure that, where a problem (such as eutrophication, acidification or susceptibility to drought) is regionally specific, measures appropriate to that particular area can be taken. The range of environmental conditions in the Community is very wide and Community policy must take this into account. The *Water Framework Directive* (see below) can be seen as one way forward in the construction of an EU-scale governance structure for water resource management at the European scale.

European Commission Legislation

Table 3.1 gives a list not comprehensive of EC legislation dealing directly and exclusively with water management issues promulgated from the 1970s onwards. The legislation is listed in chronological order,[3] together with their gazette numbers in brackets. The discussion that follows provides a more detailed discussion of the evolution of EU approaches to water management leading up to the passage of Water Framework Directive in late 2000.

Since the EU began as a customs union and political alliance it is perhaps not surprising that it was late to the enterprise of environmental regulation. However, the Stockholm conference in 1972 and the steadily growing popular consciousness of environmental issues from the late 1960s onwards meant that by the mid-1970s the (then) EEC began to turn its attention to environmental protection. The

---

3 The first two digits in the EC numerical code denotes the year of passage. So Directive 75/440/EEC, the "Surface Water Directive", was passed in 1975.

Table 3.1 EC water quality directives

Directive	Purpose/Function
The Surface Water Directive (75/440/EEC)	The first Directive designed to provide for a common surface water assessment framework. To a large extent superceded by subsequent directives.
The Bathing Water Directive (76/160/EEC)	To safeguard the health of bathers at member states' beaches (both marine and freshwater). Has been quite impactful and, indeed, popular.
The Dangerous Substances Directive (76/464/EEC)	Requires the regulation of a specified list of "dangerous substances" discharge into watercourses. Largely superceded by the *Integrated Pollution Prevention and Control Directive* (1996) and the WFD.
The Information Exchange Decision (77/795/EEC)	Marks the first attempt to create a European-level environmental database, establishing a network of monitoring points for almost two dozen different parameters. Established the groundwork for the European Environment Agency.
The Freshwater Fish Water Directive (78/659/EEC)	Requires waters designated as "fishing waters" to be kept in a state capable of maintaining fish stocks. The WFD (2000) superceded it, essentially re-creating it as a "daughter directive".
The Shellfish Water Directive (79/923/EEC)	Like the above, but for shellfish-supporting waters.
The Groundwater Directive (80/68/EEC)(36)	Requires member states to control the direct and indirect discharge of listed substances through a system of monitoring and discharge consents.
The Drinking Water Directive (80/778/EEC)	Establishes EU standards for drinking water quality. Will stand alongside the WFD (2000) as a "sister directive".
The Urban Waste Water Treatment Directive (91/271/EEC)	Requires member states to reduce nutrient (primarily nitrate and phosphate) pollution in treated wastewaters.
The Nitrates Directive (91/676/EEC)	A classic "daughter directive", it requires member states to reduce environmental pollution from nitrates to specified levels and to identify and monitor "Nitrate Sensitive Areas".

Surface Water Abstraction Directive (WFD) (75/440/EEC) of 1975 marked the Commission's first direct foray into water quality management. In keeping with subsequent Directives, and with the EC approach overall (outlined above) this Directive did not prescribe the exact manner in which the protection of surface waters should be monitored and regulated for quality but it did require that such systems be put in place by a deadline, along with Action Plans for ensuring compliance. A sister Directive (79/869/EEC) established the tripartite classification system for surface water monitoring still in use in many EU states.

The *Bathing Water Directive* (76/160/EEC) requires Member States to identify marine and fresh water bathing waters, monitor them and take "all appropriate measures" to ensure compliance with a series of water quality parameters. The Commission reports on the implementation of the Directive and on the quality of Community bathing waters every year. It is not related to the "Blue Flag" programme managed by the Foundation for Environmental Education in Europe. The Directive is a very popular one with European citizens and nobody questions its value in protecting the health of swimmers and bathers. At the time of its adoption there was little other legislation regarding the protection of waters from urban waste water and the Directive therefore had a secondary purpose of requiring Member States to take action to deal with the worst cases of pollution by urban waste water. The Thatcher government of the 1980s distinguished itself by attempting to severely circumscribe the application of the Directive by manipulating the definition of "bathing waters" (Cook, 1989). Subsequently, in the UK the BWD implementing regulations have been updated twice, in 1991 and 2003 and is now well cross-referenced in UK water quality legislation, for example the 1991 *Water Resources Act*. By 2007 there were 578 listed "bathing waters", 96.5% of which were in compliance with quality standards. Even after the WFD BWD (revised and strengthened in 2006) will remain a freestanding Directive, though actions taken to improve water quality will fall under those of the WFD.

A group of Directives promulgated in the late 1970s and early 1980s showed the EC expanding the orbit of its water management activities into more areas, and deepening them also. From the *Dangerous Substances Directive* of 1976 to the *Shellfish Water Directive* of 1979 the EC signalled its intent to create a comprehensive system for water quality management across all member states and all sectors of activity (e.g. fishing, water pollution control, etc.). The *Information Exchange Decision* for the first time created the conditions for comparable and consistent monitoring of water quality across member states, and was a necessary precursor to the European Environment Agency, established in 1990. The *Groundwater Directive* (80/68/EEC, updated in 2006) of 1980 established a system for regulation of substances entering groundwater but allowed member states to determine which substances would received priority ("List 1") treatment and which would not. Currently the Environment Agency for England lists more than 100 List 1 substances, most of which are organohalogens – substances used in agricultural pesticides and herbicides and also in industrial solvents.

With the *Drinking Water* (80/778/EEC) and the *Urban Wastewater Treatment* (91/271/EEC) Directives attention shifted towards water quality in its relations to human use. The former directive, significantly updated in 1998, established a list of more than 60 water quality parameters and a mechanism for reporting against these. The Directive is rather different to other pieces of water legislation in than it sets rigorous prescriptive standards. Failure to report against these parameters or to implement the directive can lead to prosecution in the European Court of Justice, as it did effectively against Ireland in 2000 for failure to comply with coliform limits (see Chapter 4 for more information about coliform as an indicator of water quality and also Table 3.2). In the Irish case, the EC used the court application to help force actions in Ireland to bring it into compliance with the Directive. The impact of this Directive has been significant and it is generally recognised that it has been the driving force behind the overall improvement in drinking water quality which has taken place in the Community over the past decade.

Similarly the *Urban Wastewater Treatment Directive* aims to reduce the pollution of surface waters with nutrients (particularly nitrates and phosphates) from urban waste water, one of the major sources of nutrient pollution and, hence, of eutrophication. It also has the objective of reducing nitrate concentrations in water abstracted for the provision of drinking water. The Directive is a good example of combining the use of the environmental quality objectives approach and the emission limit values approach. Related directives, including the *Nitrates Directive* (91/676/EEC), provide for more detailed regulation of specific wastewater pollutants, including nitrates, pesticides and more recently phosphates. An important new approach to managing these specific pollutants is the requirement to identify geographical areas that may be at risk of pollution levels in excess of that specified in the directive, that is, 50 mg/l (see also Table 3.2). By late 2002 approximately 3,600,000 hectares, or 55% of total land area had been designated as NVZ and there was a growing realisation that the original directive had probably been too prescriptive. For one thing the 50 mg/l limit meant that the spreading of animal manures on arable land would have to cease in many areas, thus creating a disposal problem for wastes formerly used as natural fertilisers and soil enhancers (House of Commons, 2008). In fact the debate about how exactly to implement the Nitrates Directive is very instructive in terms of the ways that member states have appealed to the implementing principles of "proportionality" and "subsidiarity" introduced in the previous section of this chapter.

As can be seen from the above, from a standing start in the early 1970s the EU has now developed an increasingly complex and comprehensive water management system. The Water Framework Directive, to which we turn our attention next, marks both the pinnacle achievement and also a radical shift in direction. Contrary to those who think that only the WFD now matters, the EC itself has made it clear that it will continue the process of refining the pre-existing Directives and introducing new ones where appropriate (e.g. for phosphates). What the WFD does in fact is to fundamentally alter the manner in which such directives can be approached (and as such it marks a significant constraint on the principles of proportionality and subsidiarity). Thus it may be considered a manifestation of the process of "deepening" the EU begun with the Single Europe Agreement of 1986 and the Maastricht Treaty of 1992 up to and including the more recent Lisbon Treaty of 2007.

Table 3.2   EC drinking water standards (Directive 80/778/EEC)

Parameter	Expressed as	Guide Level	Max. Allowable Concentration
Colour	mg/l Pt/Co	1	20
Turbidity	Kg/l Si/O$_2$	1	10
Odour	Diln #	0	2-3
Temperature	° Celsius	12	25
pH	Unit	6.5-8.5	
Conductivity	uS/cm	400	
Chloride	mg/l as Cl	25	
Sulphate	mg/l as SO$_4$	25	250
Calcium	mg/l -as Ca	100	
Magnesium	mg/l as Mg	30	50
Sodium	mg/l as Na	20	175
TDS	mg/l as Cl		1500
Nitrate	mg/l as NO$_3$	25	50
Ammonia	mg/l as NH$_4$	0.05	0.5
Phenols	mg/l as C$_6$H$_5$OH		0.5
Boron	μg/l as B	1000	
Iron	μg/l as Fe	50	200
Manganese	μg/l as Mn	20	50
Phosphorus	μg/l as BP$_2$O$_5$	400	500
Fluoride	μg/l as F	1500(12° C)	700(25° C)
Arsenic	μg/l as As		50
Cadmium	μg/l as Cd		5
Cyanide	μg/l as CN		50
Mercury	μg/l as Hg		1
Lead	μg/l as Pb		50
Pesticides	μg/l		0.5
PAH	μg/l		0.2
Total Coliform	MPN/100 ml		< 1
Faecal Coliform	MPN/100 ml		< 1
37 degree colonies	per ml	10	
22 degree colonies	per ml	100	

European Water Management for the 21st Century: The *Water Framework Directive*

The legal framework for water management policy in the Member States of the Community consists of a combination of measures derived from Community legislation and national measures. Many of these measures have been adopted almost ad hoc over the last three decades. This de facto framework was fundamentally changed in December 2000 when the EC adopted the *Water Framework Directive* as an attempt to create a unified approach to water quality management out of the plethora of specific technical water quality directives developed over the preceding 20 years. It requires all inland and coastal water bodies to reach at least "good status" (defined in chemical and biological terms) by 2015. It will do this by establishing a river basin district structure within which demanding environmental objectives will be set, including ecological targets for surface waters. The WFD therefore sets a framework which should provide substantial benefits for the long term sustainable management of water. A key element of early implementation of the WFD is the creation of an EU pilot river basin network, comprised of 15 river basin projects. The UK initially participated in this network through the Ribble pilot river basin project, located in the North West river basin district, which was formally launched 10 June 2003. From 2005 all other RBDs within England, Scotland and Wales have been integrated into the WFD-mandated planning process.

Key features of the WFD are:

- The concept of river basin management is introduced to all Member States through the establishment of "river basin districts" as the basic management units. For international rivers these river basin districts (RBDs) will transcend national boundaries (Article 3).
- For each river basin district a "river basin management plan" must be developed, including a "programme of measures", and these will form the basis for the achievement of water quality protection and improvement (Articles 11 and 13).
- Although its prime aims are environmental, the WFD embraces all three principles of sustainable development: environmental, economic and social needs must all be taken into account when river basin management plans are being developed (Article 9).
- The river basin management plans will not allow further deterioration to existing water quality. With certain defined exceptions, the aim is to achieve at least "good" status for all water bodies in each river basin district. Geographical factors are allowed for when good status is defined and the principle of "subsidiarity" allows Member States some freedom within the overall requirements of the WFD (Article 4).

- The two previously competing concepts of water quality management, the use of environmental quality standards and the use of emission limit values are brought together by the WFD in a new dual approach (Article 10).
- To overcome the previously piecemeal nature of water environment regulation, a number of existing directives will be replaced when new local standards are developed to meet WFD requirements. These local standards must be at least as stringent as those being replaced. Daughter directives will be introduced to deal with groundwater quality and for priority (dangerous) substances (Article 16).
- Measures to conserve water quantity are introduced as an essential component of environmental protection. Unless minimal, all abstractions must be authorised and, for groundwater, a balance struck between abstraction and the recharge of aquifers (Article 11).
- The polluter pays principle is incorporated through a review of measures for charging for water use, including full environmental cost recovery (Article 9).
- Public participation and the involvement of stakeholders is a key requirement of the river basin management planning process, thus satisfying this aspect of Agenda 215 (Article 14).

Whilst EU actions such as the Drinking Water Directive and the Urban Waste Water Directive can duly be considered milestones, European water policy had to address the increasing awareness of citizens and other involved parties for their water. At the same time water policy and water management are to address problems in a coherent way. This is why the new European water policy was developed in an open consultation process involving all interested parties. Yet, as Page and Kaika (2003) point out, the WFD seems also to have favoured increased privatisation of the water commons by (over)relying on the notion that water in all its facets is subject to costs and that these should be borne as much as possible to those nearest their *causus loci* (see also Kallis and Butler, 2001; Kallis and Nijkamp, 2000).

Case Study: The UK

Water policy in the UK aims to protect both public health and the environment by maintaining and improving the quality of water and its optimal apportionment between competing uses. To some extent the government's priorities are determined by European Community (EC) law and other international agreements which have been systematically adopted into UK law. There are over 300 EC Directives dealing with environmental matters; about 30 of which concern water directly or indirectly. The ten most important (plus the WFD) have been described above. Some of the principles which underlie domestic UK legislation are described in this section. As well as considering the application of these principles, the UK

Government seeks to ensure that there is a proper assessment of the costs and benefits of any action affecting the environment.

Water management in the UK has been radically reorganised three times in the Post-World War II period; first with the 1945 *Water Act*, second with the 1973 *Water Act* and, finally, with the 1989 *Water Resources Act*. The first act, in 1945 began the process of consolidating a very confused water supply system that was comprised of over 1,000 statutory water undertakers in 1945 to only 200 in 1974. The 1973 *Water Act* further consolidated the sector into 10 regional water authorities who also gained control over the myriad sewerage companies that had not been included within the 1945 *Act* (MacDonald and Kay, 1988, p. 213). One of the purposes of the 1973 Act was to bring the whole of the water cycle into the hands of a limited number of public authorities responsible for abstraction, discharge and overall management of the water environment (Maloney and Richardson, 1995). At the time some critics decried the loss of local control over water allocation, infrastructure planning and finance, although the logic of Integrated Resource Management (see Chapter 5) won the day.

The 1973 Act also contained the first powers to enforce water metering – a provision strenuously opposed by the then-opposition Labour Party (see Chapter 6 for discussion of Labour's *volte face* on the metering issue) – and created the corporate structures that would eventually lead to privatisation in 1989. Interestingly, the Thatcher governments of the early 1980s were initially quite reluctant to privatise the water sector, arguing in 1984:

> We have absolutely no intention of privatising the water industry. The government has no plans to urge that upon the water authorities. There has been some press speculation about it in the past, but there is no intention to do so (Neil MacFarlane, Parliamentary Undersecretary of State for the Department of the Environment, HC Debates, 20 December 1984, col.457).

Yet within two years the Conservative government had made a u-turn and published a 1986 White Paper entitled "Privatisation of the Water Authorities in England and Wales" (Maloney and Richardson, 1995). The 1989 *Water Act* mandated the privatisation of these 10 regional water authorities and promulgated a new basis for water services allocation – *ability to pay*. Related legislation in 1991, the *Water Industry Act* and the *Water Resources Act*, completed the new architecture of water services regulation more or less as it currently exists. Finally the *Water Industry Act* of 1999 amended the 1991 Act, curtailing industry's power to disconnect water users (even if they did not pay) and to impose metering and required private water companies to offer special payment schemes for "vulnerable groups". 2003s *Water Act* amends the 1989 *Act* by enlarging the scope for competition between private water providers.

Over the six years immediately after privatisation, as Ward (1997) points out, the water bill for the average householder has risen by approximately 67%. On the other hand there has been considerable new investment in water supply and sewerage infrastructure, and the "regulated monopoly" structure of the current water industry in England and Wales seems largely to strike a productive balance between bringing in new investment and protecting the rights of water consumers. Successive pieces of legislation over the last 15 years have clarified and strengthened the mechanisms for representing consumers' interests, a function now fulfilled by the Consumer Council for Water.

In what follows we review the primary institutions that manage England and Wales' water environment. Since devolution in 1997 Scotland has managed its own water environment and Northern Ireland is still managed from Whitehall (until the devolution process there can be completed).

In England, the Department for Environment, Food and Rural Affairs (DEFRA) oversees water policy and is the primary controller and overseer of the other key water management institutions including the Environment Agency, the Drinking Water Inspectorate, the Office of the Water Services Regulator (OFWAT) and British Waterways. DEFRA leads the Government's drive to integrate environmental concerns into decision-making. Its aims for sustainable water management are orientated towards timely, cost-effective, consultative and sustainable management of the water environment. These aims are achieved through a balance of:

- Legislation to set appropriate environmental standards, backed up by various regulators which monitor and enforce them;
- Economic instruments such as tradeable permits or charges to discourage the use of polluting substances, so that businesses and industry work for the environment; and
- Voluntary initiatives to encourage businesses and citizens to act responsibly.

The Government works with other states at regional (e.g. European Union and North Sea Conference) and global (e.g. United Nations) levels, to protect human health and improve the quality of the environment.

The *Environment Act 1995* established the Environment Agency, and introduced measures to enhance protection of the environment, including further powers for the prevention and remediation of water pollution. The Environment Agency makes sure that these policies are carried out. The Agency has a responsibility to protect and enhance the environment as a whole, monitoring and enforcing aspects of air quality and waste as well as water.

Environment Agency

The Environment Agency (formerly the National Rivers Authority) is the primary government agency involved in implementing UK and EC legislation with respect to water quality. It operates a monitoring and enforcement mechanism based on a series of staged procedures:

Protection through consents   Under the *Water Resources Act* 1991 it is an offence to cause or knowingly permit polluting matter to enter into "controlled waters", that is, rivers, estuaries, coastal waters or groundwaters without permission. Permission is generally obtained as a discharge consent granted by the Environment Agency. The Agency sets conditions which may control volumes and concentrations of particular substances or impose broader controls on the nature of the effluent. Each consent is based on the objectives set by the Agency for the quality of the stretch of water to which the discharge is made as well as any relevant standards from EC Directives. The Environment Agency may also refuse an application for a discharge consent.

The majority of trade effluent discharges are regulated through a different system. The *Water Industry Act* 1991 permits the discharge of trade effluent into the public sewers if the sewerage undertaker agrees. Water companies who have responsibilities for providing, improving, maintaining and cleansing public sewers are known as "sewerage undertakers". A discharge without the undertaker's agreement is a criminal offence. An application for the undertaker's permission must describe the nature or composition of the discharge, the maximum daily rate and the highest rate of discharge. This is essential information since the procedures are flexible and allow consents to be tailored to a particular situation. After consideration of an application, the undertaker may issue an unconditional consent, or it may set conditions as to the nature or composition, rates of discharge and as to which sewer(s) receive the discharge. The consent may also impose additional conditions as to timing, content, flow, inspection, control procedures and charges.

Protection through enforcement   If pollution occurs, including when the conditions of a consent are broken, a criminal offence has been committed. In these cases the polluter can be prosecuted, usually by the Environment Agency, and may be fined and made to clean up the pollution. See Chapter 4 for examples of quality management and enforcement.

Protection through pollution prevention   Prosecution is a last resort. A much better course is to prevent pollution in the first place. This can be achieved in a variety of ways. For example:

- Through education, publicity and guidance:
  - Advice can be given through codes of practice, guidance notes, pollution prevention awareness campaigns and site visits to farms and other operations;

- Through regulations:
  - Regulations establish controls necessary to prevent pollution. For example, regulations introduced in 1991 set minimum standards for the construction of stores for agricultural waste (silage, slurry, and agricultural fuel oil) since such substances can be highly polluting to water;

- Through notices requiring action:
  - The Environment Agency can serve notices on consent holders requiring action to be taken to prevent breach of a consent. There are also new powers which will allow the Agency to require potential polluters to carry out works to prevent pollution.

Protection through abstraction licensing In addition to preventing pollution occurring, the Environment Agency is responsible for ensuring that water resources are managed effectively, so that there is enough to meet all our needs. Almost anyone who wants to take water from a surface or underground source must obtain a licence to do so from the Agency. Conditions placed on abstraction from surface waters enable flow to be maintained at levels which the Agency considers necessary to ensure that quality objectives are achieved.

The Environment Agency is also the primary manager of surface and groundwater quality. For each stretch of a river the Environment Agency establishes a river quality objective. This reflects the uses to which the waters are put and is the basis for deciding the discharges which can be allowed. The objectives will generally be expressed in terms of quality grades (like those described below in monitoring water quality) which reflect the general health of the waters. The Secretary of State has powers under the *Water Resources Act* 1991 to set statutory objectives, giving the Government and the Environment Agency a legal duty to ensure that they are achieved. These powers have been used to establish environmental water quality standards for particular stretches of inland and other coastal waters to fulfil the requirements of European directives. Examples include bathing waters, shellfish waters, or sources for drinking water supply.

The water quality in about 36,000 km of rivers and canals in England is routinely monitored. General Quality Assessment (GQA) is the scheme used by the Environment Agency for classifying water quality. It will eventually consist of assessments based on a number of measures. The measure most widely used so far is chemical quality which is monitored annually on a three year rolling programme. Biological grading was also included in the Agency's 1995 River Quality Survey and is monitored on a five yearly basis. Assessments of aesthetics and nutrients are being considered for inclusion in the future.

The system assesses water quality on the basis of biochemical oxygen demand and concentrations of dissolved oxygen and ammonia. The GQA system assigns stretches of river and canals to one of six chemical water grades based on different degrees of quality from A for water of the very highest quality to F for water which is of bad quality. Typical BOD levels range from less than five mg/litre in clean river water to up to 100,000 mg/litre in industrial waste and up to 80,000 mg/litre in silage effluent from farms. Biological Grading measures water quality by comparing the number and diversity of macro-invertebrate (tiny animal) species at 7,000 sites against the range of those species that might be expected if the sites were in a pristine, unpolluted condition.

Table 3.3   Surface water quality in England and Wales: Percentage length in England in each GQA chemical grade

	Good		Fair		Poor	Bad
	A	**B**	**C**	**D**	**E**	**F**
1990	13.9	29.6	24.8	15.1	14.0	2.5
1998	21.2	33.3	22.2	12.1	10.4	0.8
2004	31.4	33.9	19.5	8.6	5.9	0.6

Source: Drinking Water Inspectorate, 2006.

Between 1990 and 2004 there was a net improvement in water quality in more than 50% of the monitored length of rivers and canals in England and Wales (Table 3.3). In England, the length of "good" quality watercourses rose from 44% to 65% while the length with poorer water quality has been reduced from 17% to 6.5%. By 1997, 82% of rivers in England and Wales complied with their river quality objectives. Though river quality in the ex-industrial Midlands was lower than most other regions of England, the northeast was not similarly afflicted.

Drinking Water Inspectorate

The Drinking Water Inspectorate's main task is to check that water companies in England and Wales supply water that is wholesome and complies with the statutory requirements of the Water Supply (Water Quality) Regulations. The Inspectorate carries out this function by technical audits of water companies. Technical audits consist of three parts:

1. An annual assessment, based on information provided by companies, of the quality of water supplied, compliance with sampling and other statutory requirements, and the progress made on improvement programmes;
2. Inspection of individual companies, covering not only a general check on the matters above but also checks that the sampling and analysis carried out by the companies is accurate and that it provides a reliable measure of drinking water quality; and
3. Interim checks made on particular aspects of compliance with the Regulations based on information provided periodically by the companies.

If companies are found to be in breach of the statutory requirements and the breach is not trivial the Inspectorate notifies the company that enforcement action is under consideration. In most cases the notification can be withdrawn if the company provides satisfactory evidence that effective remedial action to prevent a recurrence has been taken. In other cases legally binding undertakings to carry out appropriate improvements by a specified date are accepted from the company. If no undertaking is offered, or the terms of an undertaking are not satisfactory, a notice of the intention to make an enforcement order is served on the company. An enforcement order specifies the steps the company has to take to comply and the dates by which they must be taken.

Another major task of the Inspectorate is to investigate all incidents affecting or threatening to affect drinking water quality and to determine whether the company took the action necessary to protect consumers, return supplies to normal and prevent a recurrence and whether there was a breach of a standard and, if so, to consider enforcement action. Section 70 of the *Water Industry Act 1991* makes it a criminal offence for a water company to supply water which is unfit for human consumption. The Inspectorate can bring prosecutions in the names of either the Secretary of State for Environment, Transport and Regions or the National Assembly for Wales (referred to in this leaflet collectively as "the Authorities"). The Inspectorate will bring prosecutions if it believes that it has evidence that water unfit for human consumption was supplied, if it believes that the company does not have a defence that it took all reasonable steps and exercised all due diligence and if such a prosecution is regarded as being in the public interest. Similarly, the DWI investigate breaches of the treatment standard and monitoring requirements for *Cryptosporidium* as such breaches are also criminal offences under amending regulations made in 1999.

Table 3.4   England Midlands water quality in 2004

Parameter	Current standard	Number of tests	Number of tests not meeting standard
Aesthetic Parameters			
– colour	20 mg/l pt/Co	4921	0
– odour	3 @ 25° C	5267	1
– taste	3 @ 25° C	5247	3
Aluminium	200 μg/l	6024	6
Benzo(a)pyrene	0.01 μg/l	1740	2
Fluoride	1.5 mg/l	1695	1
Iron	200 μg/l	6809	29
Lead			
– Current standard	25 μg/l	1739	5
– Future standard	10 μg/l	1739	31
Manganese	50 μg/l	5268	3
Nickel	20 μg/l	1738	3
Nitrate	50 mg/l	2021	1
Nitrite	p.5 mg/l	2021	0
Pesticides – total	0.5 μg/l	1731	0
Pesticides – individual	0.1 μg/l	51000	2
pH	6.5-10	6963	2
Turbidity	4 NTU	6167	2

Source: DWI, 2005.

The DWI (like all other statutory bodies in England and Wales) takes a *regional* approach to water quality monitoring. In the Midlands region (depicted) there are 8.8 million people using an estimated 2.4 billion litres/day for an average of 265 litres/person/day (this per capita figures includes all uses: domestic, agricultural, industrial, etc.). Table 3.4 shows that there is, overall, a high level of compliance with published water standards with very few samples failing to meet standards. Where standards are not met the rightmost column denotes the remedial action taken by the DWI. The single case where there are a number of incidents of non-compliance involves the adoption of new standards for lead which dropped from 25 µg/l to 10 µg/l.

British Waterways

British Waterways is the custodian of 2,000 miles of the nation's historic network of canals and inland waterways, much of which is over 200 years old. Built to service the transport needs of the world's first industrial revolution in the late 18th and early 19th centuries, waterways transformed the social and economic life of communities all over Britain. In the new millennium, they can bring about a new renaissance. Today, the waterways are valued as a leisure and recreation resource for millions and are an integral part of land drainage and water distribution systems whilst still providing an environmentally friendly means of transport for coal, aggregates and other materials. Where once derelict canals were an eyesore and target for vandals, they are now becoming a focus for regeneration and development schemes as in the cities of Gloucester and Stroud in western England.

British Waterways is required, under the *Transport Act* of 1968, to maintain its canals and rivers in a safe and satisfactory condition in accordance with standards defined in the Act. To help it achieve this, British Waterways receives an annual grant from the Government. This is in addition to its income from boaters, property etc.; an income which is not sufficient on its own to cover the costs of maintaining the waterways. British Waterways' success in expanding its external income has been offset by reductions in real terms in the annual grant it receives from Government. Total income has been insufficient to meet the needs of the canal network and a backlog of maintenance, currently estimated at £260 million has built up. Of this, some £90 million is for work that poses a serious public safety risk. The Government considers that this backlog of safety-related maintenance is unacceptable.

British Waterways already uses its network to move supplies of untreated water to serve individual water companies (e.g. Bristol Water via the Gloucester and Sharpness Canal). It has also successfully carried out water transfer for its own operational needs (e.g. extracting rising groundwater from disused mines under Birmingham to maintain water levels for navigation on the Oxford Canal which has historically suffered from poor water supply).

The waterway network includes 3,200 km of canals, 4,763 bridges, 397 aqueducts, 60 tunnels, 1,549 locks, 89 reservoirs, nearly 3,000 listed structures and ancient monuments and 66 Sites of Special Scientific Interest. Much of the network is over 200 years old. British Waterways has worked hard to ensure best value for money in maintaining these special assets. Most work is already put out to competitive tender and is often undertaken in partnering arrangements (e.g. with Gloucester and Stroud Councils) to improve value for money.

OFWAT

The Water Services Regulation Authority (OFWAT) is the economic regulator of the water and sewerage industry in England and Wales. Its purpose is to use regulatory mechanisms and powers to ensure that the privatised water sector is run in an efficient, prudent and socially just manner. The primary duties of the OFWAT (as amended by the *Water Act* 2003) are to act in a way that it considers is best calculated to:

- Protect the interests of customers (wherever appropriate by promoting effective competition);
- Secure that the functions of each water and sewerage company, as specified in the 1991 Act, are properly carried out; and
- To secure that companies are able to finance their functions, in particular by securing a reasonable rate of return on their capital.

OFWAT also has powers under the *Competition Act* 1998, which came into force on 1 March 2000. The *Competition Act* prohibits companies from entering into agreements that are anti-competitive and prohibits abuse of a dominant market position. It strengthens OFWAT's powers to investigate complaints and to take action, including imposing financial penalties, where behaviour is anti-competitive.

Consumer Council for Water

Established by the 2003 *Water Act* as the successor to Watervoice, CCW champions consumers' interests in the highly complex world of water services. Since its inception in October 2005, the Consumer Council for Water has obtained over £591,176 in compensation and rebates for consumers. This follows over 6,300 complaints received by CCWater between October 2005 and April 2006. CCW has also taken an active part in the policy development and price-setting processes most recently with respect to water companies "Water Resources Strategies" and the price review for 2010-2014.

Concluding Thoughts

The UK water sector is often held up (by its fans) as a model of well-regulated privatisation of water services and indeed it is true that it does have some favourable elements. First, unlike many privatised water sectors in other countries, there is both a wide variety of private actors involved (though as of Autumn 2007 this is threatened by a fresh wave of mergers and acquisitions activity) (see Chapter 6 for more details on this issue) and a strong set of regulatory institutions. In particular both OFWAT and the Drinking Water Inspectorate regularly investigate and levy fines as sanctions for poor company performance or technical malfeasances (such as exaggerating performance on certain key indicators).[4] On the other hand it is also true that the regulated business environment has worked against UK consumers as well, not least because the system works essentially to guarantee certain levels of profit to the privatised companies (through the five year "Price Review" process). These issues will be discussed further in Chapter 6.

The UK water sector also faces challenges from a perhaps unusual quarter – it seems that many actors within the regulatory apparatus do not understand the economics of the water sector sufficiently well. Moreover it seems increasingly the case that politics is over-riding sound economics in the making of important policy decisions. One recent example involves the current government's espousal of water meters as a key instrument for promoting water conservation notwithstanding the fact that meters prove to be a very expensive way of inducing water savings. This issue will be taken up in Chapter 5.

Still it remains true that, whatever the precise causal mechanisms, the current regulatory system is associated with provisions of service to record numbers of consumers and at record high levels of quality.

Case Study: The Danube Basin Initiative

We would be remiss if we did not speak further about movements for supra-national water management emanating from the non-governmental, non-state sector. After all, it is a premise of this chapter that there are a number of important processes affecting water management, not all of which are contained within individual nation-states – I have referred to this in the preceding chapter as the "globalisation of water governance". The Water Framework Directive has already been mentioned, but there are also others including the Danube Basin Initiative.

The following case study presents a situation that is explicitly transnational in scale, and which therefore provides something of a more direct test of the Ostrom-inspired perspective introduced in Chapter 1 and the new paradigm of water governance mooted in Chapter 2. The environment of the Black Sea/Danube Basin has become significantly degraded over the past several decades. Pollution of the waters of the Black Sea and its tributaries, notably the Danube, has caused significant losses to riparian countries through reduced revenues from tourism and fisheries, loss of biodiversity, and increased water-borne diseases. Extensive studies

---

4   In November 2007 OFWAT upheld a fine of £20.3 millions against Southern Water for misreporting on some of its performance indicators. Similar fines have been levied against Thames Water and Severn Trent in recent years.

The Danube River Basin

Figure 3.2   Danube River Basin

conducted during the 1990s have shown that over-fertilisation of the water bodies by nitrogen and phosphorus discharges from municipal, industrial and agricultural sources were the most significant causes of the ecological degradation that the Black Sea and the Danube River have experienced. Additionally, catastrophic events, such as the NATO bombardment of Serbia in 1999 and the 2000 spill of 100,000 cubic meters of mine tailings from Baia Mare in Romania have significantly affected water quality in the Danube and, ultimately, the Black Sea.

On 29 June 1994 in Sofia, Bulgaria 11 of the Danube Riparian States and the European Union signed the Convention on Co-operation for the Protection and Sustainable Use of the River Danube (short title: Danube River Protection Convention-DRPC). The Convention is aimed at achieving sustainable and equitable water management in the Danube basin. The signatories have agreed:

- On "conservation, improvement and the rational use of surface and ground waters in the catchment area";
- "Control of the hazards originating from accidents involving substances hazardous to water, floods and ice-hazards";
- To "contribute to reducing the pollution loads of the Black Sea from sources in the catchment area" (Art. 2.1).

The signatories further agreed to cooperate on fundamental water management issues by taking:

all appropriate legal, administrative and technical measures to at least maintain and improve the current environment and water quality conditions of the Danube river and of the waters in its catchment area and to prevent and reduce as far as possible adverse impacts and changes occurring or likely to be caused (Art. 2.2).

The Strategic Action Plan for the Danube river basin was adopted during a meeting of Danube countries' Environment Ministers and the European Commissioner for the Environment on the 6 December 1994 in Bucharest, Romania. The Declaration of this meeting required the countries to "take the measures to implement the SAP including the necessary legal and administrative arrangements". The Strategic Action Plan (1995-2005) is the result of the first phase of the Danube Environmental programme, and especially a result of a number of studies and reports which have been carried out and prepared. In this period information was collected, problems were evaluated and defined, transboundary issues were identified, basin-wide water quality monitoring was implemented, monitoring and data management strategies were developed and a warning system for accidental pollution was established.

The "Global Environment Facility (GEF) Strategic Partnership on the Black Sea and Danube Basin" has been established with the cooperation of the World Bank (WB), the United Nations Development Programme (UNDP), the United Nations Environment Programme (UNEP) and other multilateral and bilateral financiers and basin countries. The Partnership aims to promote investments and capacity building to return the Black Sea/Danube Basin environment to its 1960s condition. The two elements of the Partnership are:

a. The WB Investment Fund for Nutrient Reduction in the Black Sea/Danube Basin to help finance investment projects in industrial and domestic wastewater treatment, wetland restoration and environmentally friendly agriculture.
b. Two UNDP/UNEP Regional Projects designed to enhance the capacity of individual riparian countries and their commissions (Black Sea Commission, Danube Commission) and improve the policy framework to address Black Sea and Danube pollution.

Another important sphere of institutional development is public participation. Most funded activities to date have assisted the development of NGOs which are addressing Danube related issues, and have promoted networking between them. In October 1994 the Danube Environmental Forum (DEF) was established, the Danube Grants Programme (DGP) for NGOs was launched in Spring 1995, the NGOs Focal Points (FPs) had several meetings during the 1995, and financial support has been provided to establish NGOs Information Centres (ICs) in the upper and lower parts of the basin. To support public participation with adequate information and to enhance networking between all interested parties, a brochure (*The Action for a Blue Danube*, January 1995) and a bulletin (*Danube Watch*, first issue in December 1994) have been released.

West (1999) suggests that while significant achievements have been made, there have been two key problems:

- The initiative has not developed into a cornerstone of Danube Basin management (unlike perhaps the Mekong Basin Commission), rather there has been much conflict and political struggle.
- Capital investment has not improved.

It might also be the case that in important respects the DBI has been superseded, not least by the Water Framework Directive itself.

Concluding Thoughts

This has been a long and complex chapter, starting with a consideration of the history of water supply and wastewater treatment and ending with two case studies of water management issues in different parts of Europe and at different spatial scales. Along the way we have looked at the emerging structure of water management emanating from the EU up to and including the Water Framework Directive (2000) which establishes two key principles for future developments:

1. A catchment basin scale approach to water management;
2. The importance of a participatory approach to allocation and other management decisions.

Our discussion of the WFD will be further developed in the next chapter, but I would like to conclude with a few comments about the more easterly parts of Europe here.

First, as the process of EU enlargement continues it is clear that the strains to the emergent EU water management system will continue to be felt. The ex-communist states of Central and Eastern Europe have far less extensively and intensively developed water supply and sewerage systems (World Bank, 1993). This means that there will have to be a dialogue about standards and their adoption in the new member states. Conversely the "old" member states are in some areas reaching the stage where further incremental improvements in water quality are likely to be disproportionately expensive and technically difficult. For example, the continued removal of nutrients from UK water courses may require the banning of phosphate additives from detergent products, with the likely knock-on price and technology implications that this unavoidably brings. In both cases there may be a case for adopting flexible emission limits as discussed by Smith, March and Jolma (1999) and others.

Second, there is a large debate to be had about the correct institutional structures for optimal provision of water services. There is no, as we shall see in Chapter 6, single delivery model, though there is a clear trend towards increased involvement of the private sector. Yet privatisation has not always worked or worked well.

Similarly with the regulatory infrastructure. The WFD requires the establishment of catchment management systems (CAMS) and specific information collection, dissemination and feedback mechanisms but it remains to be seen how national implementations will proceed, particularly with respect to the EU principles of "subsidiarity" and "proportionality".

Suggested Activities for Further Learning:

1. Write an essay about the history of water supply and sanitation in your own country. It would be especially interesting to consider the interactions between private entrepreneurs and government in the early development of the water network.
2. Simpson's sand filtration sewage treatment system was originally developed in the 1830s – to what extent is it related to modern sewage treatment techniques? This question could be addressed by examining the design of a modern sewage plant, such as Severn Trent's Minworth plant near Birmingham – it is one of Europe's largest sewage treatment plants.
3. Prepare a short summary of the achievements of Joseph Bazalgette in 19th century London.
4. Discuss issues related to the implementation of EC Directives on water quality in a particular country or countries. How have related issues such as subsidiarity intervened in the policy implementation process?
5. How do the EC water quality standards compare to those developed by other major bodies such as the World Health Organisation, or those of member states? How are they enforced?
6. The EC *Framework Directive on Water Quality* may have the effect of comprehensively restructuring water quality management in member states. Discuss some of the ways in which restructuring might happen.
7. The UK water management infrastructure rests on a complex network of institutions and legislation. Is the system "watertight"? Are there obvious gaps or omissions? How does this system compare with that of other European Union member states?

Chapter 4

Managing Scarce Water Resources: The Qualitative Dimension

Introduction

This chapter discusses the key processes which affect the quality of water resources, introduces elements of water quality modeling and presents technical, administrative, legal and economic means of water quality protection. The "training" part of the chapter presents a method for water quality assessment, illustrates (schematically and mathematically) the self-purifying processes in rivers and proposes the use of a simple water quality model for developing a water quality improvement strategy in a hypothetical river basin. At the end of the chapter there are specific "activities for learning" which might be helpful in orientating your study process. Specialist terms are usually italicised and in bold throughout the text and are defined in the Glossary contained in Appendix 1 and Appendices 2 and 3 provide additional technical material and exercises.

The quality of water resources is an important factor related to the possibility of use of water for various economic activities and determining factor on the life of aquatic organisms (flora and fauna), their species composition and the quality of the surrounding environment. At the simplest level changes in water quality are a result of two key factors: human activity and water self-purification processes. Human intervention is connected with more or less conscious delivery or removal of various pollutants, inherently connected with human activity, into the water. On the other hand, water (mainly surface water, but to a lesser extent also underground water) has the wonderful property in that it is capable of some degree of *self-purification*. Natural self-purifying processes can therefore compensate to some extent for residual anthropogenic water quality degradation. However, there is a limit above which natural processes are no longer capable of restoring the adequate quality.

Intuitively there are at least three ways of protecting water quality in water supply systems:

1. Use a pristine source.
2. Manufacture a pristine source through storage reservoirs.
3. Filter or treat non-pristine water sources.

Historically water providers tended to take a *supply-side* perspective, always seeking to solve quantitative or qualitative shortage issues by extending their supply system into remaining pristine sources (strategy 1), but eventually they have been forced to enact strategies 2 and 3. Some commentators now believe that the time has long since passed when such supply side strategies were appropriate and that *demand-side* management (forcing/encouraging water users to use water more efficiently) is required, as we shall see in the following chapters.

Sources of Pollution

The term *sources of pollution* is most often used to define anthropogenic inflows. These inflows may be intentional (e.g. of more or less treated effluents) or accidental, resulting from other activities linked with economic activities (agricultural runoffs, spillages or run-off from roadways, etc.). In terms of their inherent spatiality water pollution sources may be split into *point* and *non-point (or diffuse) sources*. Comparison of the share of non-point and point sources in water pollution in the US demonstrates the high degree of variability of pollution by source. Heavy metals such as cadmium and mercury tend to enter the environment primarily through diffuse mechanisms whereas organic pollution is usually localisable, that is traceable to a source.

Point Source Pollution

Point sources are those where pollution discharge occurs at a single, usually clearly identifiable, place. Industrial and wastewater discharges are common point sources of pollution. *Wastewater* is the water used for human economic and domestic needs. In terms of its origin, wastewater may be divided into domestic wastewater, resulting immediately from activities associated with daily life (runoffs from bathrooms, kitchen, washing and cleaning household premises), industrial wastes (originating from industrial plants and services, including technological water, cooling water and mining water), rainwater (running off during rainfalls or melting snow in towns, rinsing away the pollution deposited in those areas) and agricultural runoff (resulting from intensive cattle and pig breeding).

In the case of underground water, that is to say *groundwater*, point sources of pollution are associated with inadequate operation of the security containment systems of facilities ranging from large industrial plants to solid waste landfills, underground petrol storage tanks and domestic septic systems. Urbanisation also results in a certain amount of groundwater contamination. For example groundwater quality in the Marlborough and Berkshire Downs and the Kennet Valley in southern England is monitored using a partially completed network of public supply and private abstraction boreholes. The area is predominantly rural with Reading and Newbury being the main industrial and urban centres. High nitrate levels are experienced in parts of the region. The catchment around Ogbourne has been designated as both a Nitrate Sensitive Area (NSA) and a Nitrate Vulnerable Zone (NVZ) and an NVZ has also been designated at Compton.

Comment

For a long time underground waters were considered the best-protected water resources, but protection ensured by soil cover and rock formations is no longer sufficient. Pollution and contamination of underground water are dangerous in that often they are detected with significant delay in respect of the pollution penetration time and at considerable distances from the original source of the pollution. Moreover the possibility of removing the pollutants and reclamation of underground water is limited: removal of pollutants is difficult, time-consuming and expensive. Today, treatment of polluted groundwater often relies on dilution with "clean" supplies.

Boreholes in the vicinity of Reading yield groundwater chemistry indicative of urban groundwater contamination. Insecure insulating layers or improper rainwater and leachate drainage system facilitate infiltration of pollution into underground water. Therefore possible reduction of pollution will include resealing drainage systems at the waste dumps and the careful choice of dumping site which should be located in areas built of impermeable subsurface geologic formations (the thicker the stratum of these formations the better) and where underground water occurs at large depths. Where inadequate containment occurs biodegradation together with the percolation of the pollution downwards can create toxic *leachates* that form persistent *plumes* underground. Further exploration of groundwater pollution is beyond the scope of this book but for general information about groundwater management see Ashworth (2002) and Glennon (2002).

Non-point Source Pollution

Non-point (dispersed or diffuse) sources – include a group of (most often fairly small) point sources located within an area or inflow of pollution from a specified (large) surface area; for instance, agricultural runoff, polluted precipitation (important because of the large scale of this phenomenon in certain regions), and surface runoff during torrential rain. In the case of underground water excessive exploitation may also cause quality problems. Any change in the hydrodynamic regime in the exploited aquifer may cause intrusion of deeper underground water of poorer quality or, in coastal areas, of saline water. Non-point sources are very difficult, or sometimes even impossible to control. For a long time their share in water pollution was unnoticed and underestimated (mainly due to dispersion of the pollution discharge and diversity of substances and complexity of their transfer mechanisms). Potential for reduction of non-point pollution depends on its nature, e.g. reduction of loads from precipitation may be achieved by reduction of emissions into the atmosphere; loads from agriculture may be reduced by suitable agrotechnical activities (e.g. choice of suitable fertilisers, doses and application times). In the case of small dispersed point sources the problem may be solved by the construction of a wastewater system and combined WWTP or construction of household treatment wastewater facilities.

Comment

Acid Rain: In highly developed parts of Europe and North America (McCormick, 1997) mean pH values in precipitation can be as low as 4.0-4.5 (acid rains). In particularly bad cases it can be as low as 3.0 (Ebert, 1988). This significant acidity of precipitation is caused by sulphur and nitrogen oxides present in the air (oxides originating primarily from combustion of solid and liquid fuels), which react with precipitation to form sulfuric and nitric acid solutions. In inland water environments these acid rains result in increased acidity of surface water which in turn causes changed ecosystem* species composition ("a group of interacting organisms that live in a particular habitat and form an ecological community" – see Glossary). In the soil acidity causes increased mobility of pollution comprised therein, including heavy metals which then migrate into underground water. Such precipitation may also contain heavy metals (emissions from metalworks and transport), dust (production of cement, combustion of crude oil derivatives, transport), particles of sprayed fertilisers and plant protection agents (emissions from agriculture and forestry). For technical studies of ecosystem effects of elevated surface water acidity in Latvia see Cimdins and Klavins (1998) and Cimdins et al. (1995).

* Also referred to sometimes as "biocenosis", though for present purposes we will adopt the term "ecosystem" instead.

In terms of distribution in time of the inflow into the recipient waters, sources of pollution may be divided into *continuous* (e.g. discharge of municipal wastewater), *periodical* (e.g. discharge from campaign based industry) or *incidental* (due to failure: e.g. transport collision, leakage from an industrial plant or pipeline).

Currently the EU and member states are looking closely at strategies for managing phosphates in surface and subsurface waters. Whilst the two primary sources of phosphates are well-known (detergents and fertilisers), removing or reducing them is much more difficult. Several member states have banned phosphates in detergents, but others argue that the cost of the measure is disproportional to the potential benefit, whilst in others (such as the UK) the contribution from agriculture or even from sewage effluent is more significant and more difficult to manage. Some member states, including Germany, Italy, Hungary and the Netherlands have phased out phosphorus-based detergent activators and have seen gradual recovery of surface water quality (Glennie et al., 2002).

Self-purification Processes in Surface Water

We have already mentioned that natural *self-purification* processes occur in ground and surface waters, which can lead to improvements of its quality. Much surface water has a relatively high potential for self-purification, underground water potential is generally much lower owing to slower recycling rates. We shall now take a closer look at the processes which cause self-purification of surface water. Self-purification of surface water is the complete set of biological, physical, chemical and hydrological processes leading to reduction of water pollution in water courses and reservoirs and includes:

- *Biodegradation* of organic compounds, i.e. transformation of organic compounds into more simple, inorganic compounds, with participation of microorganisms; it takes place in the presence of oxygen dissolved in water and in anaerobic conditions;
- *Sedimentation* of suspended solids, i.e. slow precipitation of substances suspended in water and deposit on the bottom – takes place at locations of slow flow velocity;
- *Re-aeration* through the circulation/turbulence of water flow around obstacles;
- *Dilution* with other waters coming in from tributaries;
- *Adsorption*, i.e. retaining substances dissolved or suspended in water at the solid/liquid phase interface, e.g. on the bottom or shore surface, on plants.

The microorganisms which participate in the self-purification process consist primarily of bacteria: aerobes (requiring the presence of oxygen dissolved in water) and anaerobes (they do not occur in the presence of oxygen, but take oxygen from other chemical compounds). In aerobic conditions decomposable organic compounds (carbohydrates, fats and proteins) give the following final products: carbon dioxide, water, nitrates, phosphates. In anaerobic conditions the following are produced: methane, ammonia, hydrogen sulphite, alcohols and organic acids. The aerobic decomposition of pollution is a much more intensive and beneficially productive process which does not produce as many toxic decomposition products (though too much of anything can be toxic in certain circumstances). Most often decomposition of organic substances in aerobic conditions takes place in rivers (provided the pollution does not exceed the self-purification potential of the river) and in near-surface layers of lake or reservoir water. Intensity of decomposition depends on water temperature, accessibility of oxygen, concentration of pollutants and the presence of other toxic compounds. The latter may check the growth of living organisms in water and hence eliminate the process of biological decomposition of pollution. The oxygen required for the process comes from atmospheric oxygen dissolution in water, promoted by turbulent flow, water flow through falls, wave motion and photosynthesis by algae and water plants.

Processes which take place in underground water are dominated by physical and chemical processes, and in particular the share of aerobic processes is much smaller here. Biological processes associated with the decomposition of organic substances take place mainly in the soil stratum and in the vicinity of pollution focus. Chemosynthetic bacteria reducing the amounts of iron and manganese in water can also important.[1] Physical and chemical processes occurring in

---

1 Treatment of these processes is beyond the scope of this book, but chemosynthetic processes involve the production of energy out of carbon-based molecules in locations where natural light is unavailable and photosynthesis is therefore impossible.

underground water include: chemical processes when ions present in water enter into mutual reaction which may slow down (or accelerate) migration of pollution, neutralisation (buffering) of acid or alkaline reaction of pollutants, precipitation (or dissolution) of compounds, sorption and ionic exchange which may affect the concentration of some cations (including heavy metal cations).

Although it is obvious that direct measurement is the optimal method of collecting data on water resource quality, it is not always sufficient to explain processes which cause water quality modifications. For better understanding of the relations which occur *between* the water quality parameters, forecasting future quality levels, assessment of the impact of investment projects on water quality we use *water quality models*. These models allow analysis, extrapolation and forecasting of water quality (De Smedt, 1989). See Exercise 1 in Appendix 3 for a practical example that you can work through.

A further natural pathway for surface water purification involves infiltration of these waters downwards through sub-surface rock strata into underlying aquifers. Aquifers can be small or large and with flow through periods measured in days, years or even centuries, but whatever precise local conditions movement of water through the subsurface entails a two-way transfer of substances – a given unit of water could well lose pollutants (since certain geological strata can act as very efficient filters), but it could also gain pollutants as well. One of the most common pollutants thus gained is total hardness (related primarily to calcium and magnesium bicarbonates). This is why, in some parts of Europe where drinking water is collected primarily through groundwater pumping, water can leave a chalky-white deposit (called "scale") in kettles and around the lips of spigots. On the whole however, and with the important exception of local areas anthropogenically polluted by underground "plumes" of contamination (e.g. around old industrial sites or around modern petrol stations), the passage of water through aquifers has a largely purifying and therefore beneficial function. About 1/3 of England and Wales' drinking water comes from groundwater pumping – the supply of drinking water via groundwater sources is something discussed further in the next chapter.

Technical Measures (Tools) for Water Quality Protection

Ensuring suitable quality of surface and underground water (protection of water resource quality) is one of the major tasks of water management. The most effective method of protection of water resource quality consists in preventing water degradation in the first place by reducing the inflowing pollution loads (prevention rather than pollution; what the EC calls the "Preventative Principle"). Implementation of water resource quality protection tasks includes *efficient functioning of the monitoring system*, and *reduction of the volume of pollution discharged into the water* by application of specific technical solutions.

Efficient functioning of the *monitoring system*, which is focused on providing information on the water quality, sources of pollution and checking compliance with water protection regulations. The results of monitoring surveys allow for the forecasting of changes in water quality over time and space (as in the simple exercise above) and

> *A monitoring system involves more than mere assessment; it is the collection of a set of indicators at set locations and intervals to provide trend and profile information useful in management.*

defining the nature and scope of protective actions as well as development of water quality improvement programmes. In Chapter 3 we discussed at some length UK systems and institutions for monitoring and managing surface water quality and in particular the roles of the *Environment Agency* and the *Drinking Water Inspectorate*. In other jurisdictions, such as the US, there may be both federal and state agencies for managing water quality in its diverse uses (Cech, 2005).

It is important to remember that there is no universal baseline for "natural" or "background" water quality, because all watercourses are different. For example, a river running over chalk/limestone will have naturally higher pH than one flowing over granite (which may contain more metals than the first example). Bed width, shape, bankside vegetation and the speed of the water also all affect the watercourse's basic characteristics as well as its ability to negate pollution. Moreover, there are marked regional and national variations in water chemistry. Even rainfall varies in its chemical composition; inland rainfall might contain from 0.2-1 mg/l of sodium, while marine rainfall concentrations are up to 5 mg/l (as a result of the cycling of salts out of sea water). Similarly, trace metals concentrations can vary widely owing to different concentrations in geological parent materials, and to differential abilities of different combinations of Water Quality (WQ) parameters to "free" them. Possibly the most important metal in this case is mercury, which can be released from parent bedrock as dangerous methyl mercury through changes in the physical qualities of overlaying water (e.g. volume, mass, etc.). This is a key criticism of big dam projects (see McCully, 1997; World Commission on Dams, 2000). For all these reason, legal limits for pollutants (MPCs) are set well below concentrations known to cause harm in humans or local environments.

Reduction of the volume of pollution discharged into the water by application of technical solutions includes:

i. Tools for reducing pollution at source, prior to its discharge into the environment (*wastewater treatment plants – WWTP*),
ii. Actions for reduction of non-point pollution running off into water, and
iii. Action for limitation of pollution generation.

The third group of activities includes introduction of advanced water-effective (reduction of produced wastewater) and clean (reduction of pollution loads in wastewater) production technologies in industry and application of systems preventing migration of pollution into underground water near potential pollution focuses. Among the simplest physical means for reduction of non-point source pollution migrating into water we should mention *girdling ditches* and *buffer (protective) zones* around lakes and rivers. Protective zones are sites located on the shores, excluded from agricultural use (or development), covered by meadow vegetation, bushes or trees designed for retaining pollution surface runoff or pollution transported by shallow groundwater. In river systems this role may be played by flood plains; for lakes, apart from the buffer zone the bank vegetation is very important – it is capable of taking up considerable amounts of nutrients. Removal of bank vegetation biomass (e.g. shrubs and grasses) could cause a significant increase in nutrient transport resulting in accelerated eutrophication. Girdling ditches and channels which drive surface runoffs to the WWTP are applied rather with small lakes or parts of their shores. In the case of small point sources pollution discharge may be limited by the construction of on-site wastewater disposal systems and combined WWTP or household wastewater treatment facilities.

Reduction of loads discharged from point sources into surface water is effected by the use of various types of wastewater treatment prior to discharge. As we have seen in the preceding chapter state water management authorities frequently make a certain amount of wastewater treatment a condition of granting discharge permits. WWTP are hydrotechnical facilities focused on the reduction of pollution loads comprised in wastewater to a level which ensures safe discharge into recipient water bodies. Depending on the technology applied we may distinguish the following types of WWTP:

- *Mechanical*, where larger easily removable solids, suspended matter, oils and fats are removed; mechanical treatment comprises the following processes: screening on grids and sieves to remove larger solids, sedimentation in grit chambers and settling tanks to remove easily deposited pollutants, flotation in oil and grease traps and flotation chambers to separate grease, oils, petrols and organic solvents and filtration to remove fine pollutants which were not separated during sedimentation and flotation;
- *Biological*, where biological decomposition processes are used (the same as in self-purification of water) at sufficient oxygen supply; microorganisms causing biodegradation of pollution are suspended in the wastewater mass (activated sludge) or are fixed to a solid grain material and sprinkled with wastewater (biological bed); the decomposition by activated sludge is followed by clarification in settling tanks; biological treatment is usually preceded by mechanical treatment; biological WWTP may additionally be equipped with a module for enhanced reduction of nitrogen compounds (denitrification module, where specialised bacteria groups convert nitrates generated in the process of mineralisation of pollutants into ammonia or free nitrogen); the last step should consist of disinfection of effluent to remove pathogens;

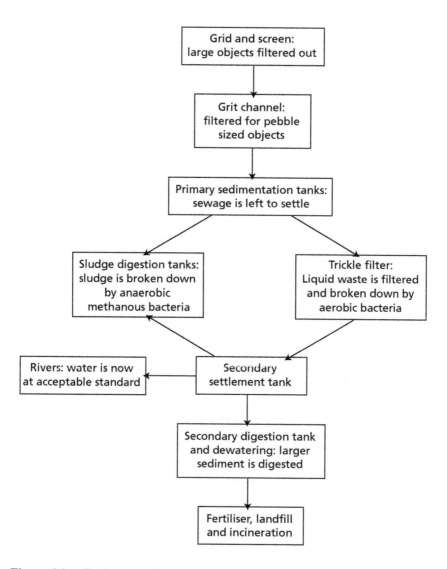

Figure 4.1   Basic stages in wastewater treatment
Source: Author.

- *Chemical,* using physical and chemical processes to drive the further separation of specific substances, oxidisation or reduction of wastewater substances (Leszczynska et. al., 1995 – e.g. heavy metals and phosphorus compounds), coagulation allowing the removal or suspended matter and colloids, neutralisation of wastewater reaction, extraction consisting in adding solvents to insoluble wastes to dissolve such substances which we want to remove from the wastewater (used e.g. to remove phenols), electrolysis by DC (e.g. to remove chromium), chemical oxidation with the use of chlorine or ozone (e.g. removal of phenols, humus acids, compounds generating colour and odour), ion exchange consisting in the use of special substances capable of exchanging their ions for ions comprised in the effluent (e.g. removal of copper, chromium, zinc, dissolved salts).

Municipal wastewater (which most often consists of domestic wastewater with the admixture of stormwater and pre-treated industrial wastes) is usually treated with mechanical and biological or biological with improved removal of specified nutrients techniques. The most common nutrients requiring specific tertiary treatment (often with expensive flocculation or UV treatments) are nitrates and phosphates. Figure 4.2 presents the percentage of population of chosen countries connected to municipal WWTP. In some cases biological treatment may also be insufficient due to the amounts of phosphorus and nitrogen left in the effluent; in such cases it is necessary to extend conventional biological treatment onto a phase for removal of nitrogen compounds (denitrification) and/or chemical precipitation of phosphorus compounds. In the case of relatively small sources of domestic wastewater specially prepared facilities such as filtration beds, sprinkled slopes, algae and oxidation ponds and bogs are used. Technical solutions differ but the pollutant reduction principle is always the same: they are based on biochemical decomposition. For the most part the majority of municipal WWTPs now incorporate primary (mechanical) and secondary (biochemical) treatment methods, usually with specific add-on treatment modules to reflect local water quality issues.

In the UK the vast majority of the population is connected to at least secondary wastewater treatment (Figure 4.2) and indeed the proportion connected to secondary and tertiary treatment is relatively high too, compared with the rest of Europe. At Minworth Treatment Works, near Birmingham, England, more than 500 megalitres of effluent are treated every day to at least advanced secondary standard with a recently greatly extended capacity for activated sludge treatment. Tertiary treatment generally involves treatment of wastewaters to a higher overall standard as well as removal of specific targeted pollutants such as nitrates, ammonia and phosphates as well as disinfection through UV light exposure (Photo 4.1). At this stage treatment methods can be very high tech and, indeed, expensive. For example at this stage wastewaters or sometimes exposed to ultraviolet light in order to aid in the oxidisation of pollutants like nitrate and phosphate.

Reverse osmosis techniques for removing phosphorus (in cases required by EU or national environmental regulations), is both expensive in energy intensive. Direct chemical treatment of wastewaters is increasingly less popular because of the need to subsequently remove the chemical reagents.

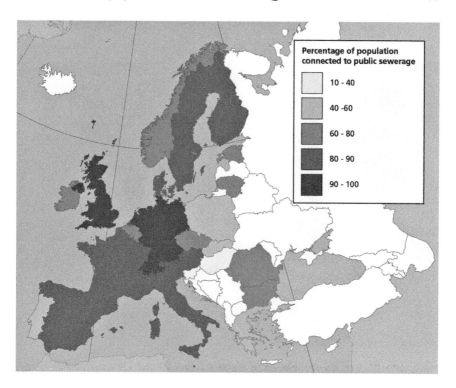

Figure 4.2   Proportion of population connected to at least secondary WWTP in selected European countries, 2002

Managing Water Quality: A Brief Introduction to a Very Technical Topic

Water quality parameters are generally set by statute, though they may be set somewhat differently by different levels of government and, as we saw in Chapter 3, there can be some confusion comparing the standards set by different governing bodies operating at different spatial scales. In the UK the *Environmental Protection Act* (1990) regulates 370 substances emitted into air, water and land. Permits, called *discharge consent limits*, to emit listed pollutants are regulated under the *Water Resources Act* (1991, revised 2003). As we saw in the last chapter, these UK national laws are in the process of being harmonised with the new EU *Water Framework Directive* (2000). Of course, water quality monitoring for different purposes (e.g. recreation versus drinking) will involve different portfolios of WQ indicators.

Photo 4.1   Tertiary treatment of wastewater

Source: Author.

Common Techniques for Measuring Water Quality

Water quality analyses are done by several methods. The most common types of measurements are *gravimetric* (weighing), *electrochemical* (using meters with *electrodes*) and *optical* (including visual). Instrumental methods are becoming increasingly popular, and instrumentation is getting "smarter" and easier to use with the inclusion of microprocessors and integrated data logging facilities. In the simplest case, a water sample of unknown quality may just be placed in an instrument and a result read directly on a display. More often some physical separation technique or chemical procedure is needed before a measurement is made, in order to remove interferences and transform the *analyte* – the target of the analysis – into a form which can be detected by the instrument. Most commonly this involves adding a chemical reagent which reacts with a specific water quality parameter of interest (e.g. nitrates) and then assessing the resultant colour of the sample in terms of its relative intensity (generally, the darker the colour the more of the given substance is present in the sample).

It should be noted that not all WQ indicators are "spot" indicators, referring to *instantaneous* concentrations. Some statutory limits are averaged over longer periods of time; an hour, a day or even a week. Biological Oxygen Demand, for example,

Photo 4.2   Colourmetric testing for nitrates

Source: Author.

is generally reported over a 5 day period – BOD_5. Sometimes this means that tables reporting WQ data will include both spot concentrations and periodic averages (the parameters and results reported in Tables 3.2 and 3.4 are spot indicators).

Since even raw sewage is generally more than 99.9% water, most environmental analyses are measuring very low concentrations of materials. The results of these measurements are usually expressed in the units "milligrams per litre," abbreviated as mg/l (refer again to Table 3.2). Since a milligram is one thousandth of a gram, and a litre of water weighs a thousand grams,[2] 1 mg/l is approximately equal to *one part per million* by weight. A part per million ("ppm") is only one ten thousandth of one percent (0.0001%), so saying that there is 1 ppm of a substance in a sample of water means that the sample contains 0.0001% of that substance; a very small amount indeed. For toxic metals and organic compounds of industrial origin, measurements are now routinely made in the part per *billion* (microgram per litre) range or even lower. At such low levels, sensitive equipment and careful technique are clearly essential for accurate results. Avoiding contamination of the sample and using methods which prevent interferences from other substances in the water are crucial requirements for successful analyses.

---

2   This is in fact the definition of a litre – a volume of water that weighs 1 kilogram.

In addition to the range of physical and chemical parameters noted above, it is also important in many cases to consider the background biological health of the water body or source. Biological health of a watercourse is defined in terms of the number and diversity of aquatic life present. Special attention is usually paid to the relative presence or absence of species that are indicators of water quality such as Caddis Fly, Stone Fly, or Bloodworm.[3] Based on these measures the UK Environment Agency (and its European analogues) has developed a general schema for classifying biological water quality.

As a rule of thumb, presence or absence of molluscs, such as mussels and clams are very good indicators of high water quality as they are quite sensitive and will not survive even modest deviation from natural (unpolluted) conditions. Even variations in water temperature (caused, say, by the use of water for industrial cooling purposes) can decimate mollusc populations. Beyond this there is a sliding scale of biological markers, from Caddis Fly and Stone Fly (good) to Bloodworms and Leeches (poor) whose relative presence gives a good indication of overall water quality.

There are several ways of assessing biological health of watercourses. One of the most common, in the UK at least, is the Biological Monitoring Working Party (BMWP) system. The BMWP was published in 1980 for a national survey of water by the National Water Council (NWC). It scores certain types of invertebrates depending on the conditions that suit that particular species. In the BMWP Score system, 82 different groups of animals are given scores that represent their tolerance to pollution. Animals that are intolerant to pollution are given a high score and those that are tolerant to pollution are given a low score. When a sample of animals is collected from the river the scores of all the

Table 4.1   Biological water quality classification: Based on Environment Agency (1998)

Grade	Outline Description
A – Very Good	Biology similar (or better than) that expected for an average and unpolluted river of same type and location. High diversity of taxa with dominance by any single taxa rare.
B – Good	Biology falls a little short of that expected for an unpolluted river. Small reduction in the relative number of taxa that are especially sensitive. Hoglouse and Mayfly could be dominant.
C – Fairly Good	Biology worse than expected for unpolluted watercourse. Many sensitive taxa absent entirely or greatly reduced.
D – Fair	Sensitive taxa scarce, pollution-tolerant taxa dominant.
E – Poor	Pollution-tolerant taxa only. Worms, leeches, maggots dominant.
F – Bad	Only highly tolerant taxa present, such as worms, leeches, etc.

3   The UK Centre for Ecology and Hydrology – http://schools.ceh.ac.uk/advanced/ freshpoll/freshpoll2.htm provides much useful information for the layperson about invertebrates used for classifying water quality.

different groups of animals present are added together to give the site score. If there are many pollution intolerant groups present then the site score is high and its biological condition tends to be good. When this score has been calculated, according to a rigorous formula, it is possible to determine how polluted the water body is by organic pollution. Because different types of watercourse can support different ranges of animals, the software package RIVPACS (River Invertebrate Prediction and Classification System) was developed to predict the taxon richness and Average Score Per Taxon (ASPT) to be expected at each different sort of site, if those sites were unpolluted. The expected values for a particular site may be considered to be its reference state. The true biological condition of sites can be judged by comparing the actual observed values of taxon richness and ASPT at a site with those that are expected if the site is not polluted.

Microbiological pollution of water is usually equated with sewage, or faecal coliform pollution. The difficulty here is that there is no single microbiological indicator here, rather there is a wide variety of associated coliforms, and different water management regions often use different coliforms as indicators of microbiological pollution. Two common coliform measures are E. Coli and Total Coliform (McDonald and Kay, 1998). Streptococci is also used. Surprisingly one still encounters disagreement in some scientific and policy circles about the danger to public health posed by contact with microbiologically polluted water. A 1959 UK study, that is still cited (!), concluded that:

> Bathing in sewage polluted sea water carries only a negligible risk to health, even on beaches that are aesthetically very unsatisfactory...the minimal risk attending such bathing is probably associated with chance contact with intact aggregates of faecal material that happen to have come from infected persons... a serious risk of contacting disease through bathing in sewage polluted sea water is probably not incurred unless the water is so fouled as to be aesthetically revolting (quoted in McDonald and Kay, 1998, p. 143).

Other studies elsewhere have suggested links between contact and disease-risk, but the debate about scope and scale rages still. In the 1980s the UK government sought to circumvent the EC Bathing Water Directive by defining "bathing waters" in restrictive terms – even Blackpool escaped designation as a bathing beach!

Evaluation of water quality by measuring faecal coliform is a standard technique undertaken in major river systems and along the UK's coastline. There are over 400 beaches subject to regular (roughly weekly) faecal coliform assessment. This type of assessment differs from others inasmuch as results are not instant (by whichever technology you are using), but rather are produced over 12-18 hours through growing visible coliform communities in a special medium. To meet EC standards samples must meet the following standards: total coliforms less than or equal to 500 coliforms per 100ml; less than 100 faecal coliforms per 100ml; less than 100 faecal streptococci per 100ml. For drinking water these are specified in EU Directive 98/83/EC "Quality of Water Fit of Human Consumption".

Common Measures of Physical and Chemical WQ

There are literally *thousands* of possible parameters that can be used to assess water quality, many of which are used in particular contexts only where particular types of pollution are anticipated. Below I present only a brief introduction to those most commonly used chemical and physical parameters, together with some comments about how to interpret them. See Appendix 2 for further information.

The scale which is used to describe the concentration of acid or base is known as "pH", for **p**ower or **p**otential of the **H**ydrogen ion. A pH of 7 is neutral. pHs above 7 are alkaline (basic); below 7 are acidic. Within a pH range of between 5.5 and 8.5 is considered tolerable for most aquatic life. The scale runs from about zero, which is very acidic, to fourteen, which is highly alkaline and is *logarithmic*, meaning that each change of one unit of pH represents a factor of 10 change in concentration of hydrogen ion. So a solution which has a pH of 3 contains 10 times as many (H^+) ions as the same volume of a solution with a pH of 4, 100 times as

Table 4.2   Common water quality parameters

WQ Parameter	Interpretation
pH	The scale used to describe the concentration of acid or base is known as pH, for power/potential of the Hydrogen ion. A pH of 7 is neutral whilst pH's above 7 are alkaline (basic) and below 7, acidic.
Dissolved oxygen	Dissolved oxygen is one of the best indicators of general water quality. As a general rule, the higher the DO, the better the water quality. DO between 6 and 15 ppm is considered desirable.
Alkalinity	Alkalinity is a measure of all of the substances in water which have the ability to react with the acids in water and "buffer" acidity, much as a Tums or Alka Seltzer does in your stomach. Water with a alkalinity of 100-120 ppm is considered to be the best for fish and aquatic organisms.
Carbon dioxide	If carbon dioxide levels are high and dissolved oxygen levels are low, fish have trouble carrying on respiration. 6 to 12 ppm CO_2 is considered good for most fish.
Hardness	Hardness refers to the amount of calcium and magnesium in the water. The natural source of hardness is usually limestone rock. Drinking water with a total hardness of 250 ppm is best. Over 500 ppm can make you ill.
Nitrates/nitrites	The presence of excessive amounts of nitrogen (nitrate/nitrite) compounds in water supplies presents a major pollution problem and can be harmful to humans. Nitrates in conjunction with phosphates can cause algal blooms and resultant eutrophication.
Phosphates	Phosphates in water stimulates the growth of algae. This in turn can lead to accelerated eutrophication of a body of water. The concentration of phosphates in water is normally not more than 0.1 ppm unless polluted.

many as one with a pH of 5, a thousand times as many as one of pH 6, and so on. To help put this technical measure in context, carbonated beverages are relatively acidic (pH between 2 and 4), vinegar is about pH 3 and lemon juice about 2.3. Common alkaline materials include baking soda (about pH 8.4) and lye (pH 14).

Dissolved oxygen is one of the best indicators of general water quality. As a general rule, the higher the DO, the better the water quality. If possible, the dissolved oxygen test should be done on site. Dissolved oxygen is also dependent on temperature: warm liquids hold less dissolved oxygen than cold liquids. Also, when organic wastes decompose in a body of water, dissolved oxygen is used up. Because more aquatic organisms are cold blooded, their metabolism rises as temperature goes up and the amount of available oxygen goes down. This often results in fish "kills", especially if the DO drops below 5 ppm.

Like solids and liquids, gases can dissolve in water. And, like solids and liquids, different gases vary greatly in their solubilities, i.e. how much can dissolve in water. A solution containing the maximum concentration that the water can hold is said to be *saturated*. Oxygen gas, the element which exists in the form of O_2 molecules, is not very water soluble. A saturated solution at room temperature and normal pressure contains only about 9 parts per million of DO by weight (9 mg/L). Lower temperatures or higher pressures increase the solubility, and *vice versa*.

Hardness in water refers to the amount of calcium and magnesium in the water. The most common natural source of hardness is usually limestone rock. The most frequent test performed on water is for total hardness. Hardness is important to living organisms because soft water makes heavy metals such as mercury and lead more poisonous to fish. Some nonmetals such as ammonia and certain acids are also more toxic to fish in soft water. There is some evidence that humans who drink soft water over a long period of time are more susceptible to cardiovascular disease. Vertebrates need calcium to build bones. All living things need calcium and magnesium in order for proper cell functions. Current public health advice suggests that drinking water with a total hardness of around 250 ppm is best. Over 500 ppm can make you very ill.

Nitrogen compounds are essential for healthy plant growth. Nitrogen is a major constituent of commercial fertiliser. The presence of excessive amounts of nitrogen compounds in water supplies presents a major pollution problem. Large amounts of nitrates and nitrites in water are harmful to humans. Nitrates in conjunction with phosphates can cause algal blooms. The US EPA states that 10ppm of nitrate/nitrogen is a limit that should not be exceeded. Lower amounts are desirable, but in some areas of the UK nitrate concentration in water is in the region of 100 ppm.

Nitrites, which are more toxic, are present primarily as the result of microbial reduction of nitrates. Associations between nitrates and cancer have been extensively investigated (McDonald and Kay, 1988).

Other fairly common WQ parameters include:

- Suspended solids (dry weight, mg/L) and/or Turbidity
- Chloride (mg/L)
- Sodium – dissolved and total (mg/L)
- Potassium – dissolved and total (mg/L)
- Magnesium – dissolved and total (mg/L)
- Aluminium – total and labile (µg/L)
- Tin – dissolved and total (µg/L)
- Manganese – dissolved and total (µg/L)
- Iron – dissolved and total (µg/L)
- Vanadium – dissolved and total (µg/L)
- Nickel – dissolved and total (µg/L)
- Mercury – dissolved and total (µg/L)
- Copper – dissolved and total (µg/L)
- Zinc – dissolved and total (µg/L)
- Cadmium – dissolved and total (µg/L)
- Lead – dissolved and total (µg/L)
- Arsenic – total (µg/L)

Another, more complex, form of chemical pollution involves the class of "organic" chemicals such as benzene, heptachlor and vinyl chloride. This carbon-based class of chemicals is very widely used in industry, and can also synthesise spontaneously when two or more constituent elements are present in a water source simultaneously – making their management an especially difficult business. Consequently there has been considerable research into their effects on human health and on the precise nature of their transit through surface and ground water systems. As McDonald and Kay (1988) point out, the co-presence of multiple organic compounds in a given water supply makes the interaction and dose-response relationships in humans quite difficult. A large proportion of the organic chemical loading in watercourses is caused by pesticide residue, including such dangerous compounds as 2,4,-D, Cyanazine and DDT. Studies in US Midwestern cities have found up to nine different pesticide residues in drinking water (Lewis, 1996).

If you look back to Table 3.2 which describes EC drinking waters standards as laid out in the Drinking Water Directive you will see most of the above parameters listed, together with guideline limits and absolute limits. In all cases the guideline limits are significantly lower than the maximum allowable concentrations, and this follows the global trend in managing water quality. The continual improvement of our scientific understanding of the environmental and health impacts of water pollution is driving down maximum allowable concentrations, in some cases towards zero. For human-made compounds, like the organohalines discussed in the previous chapter this is to be expected. Where it sometimes becomes controversial with pollutants that have natural as well as artificial sources, such as the eutrophication agents nitrates and phosphates.

Acidification

Increased concentration of acids in water bodies and in precipitation was first observed in the 19th century, but it was not until the mid-20th century that the cumulative effects of more than a century of heavy industrialisation made the problem sufficiently serious and sufficiently widespread to raise alarms in scientific, policy and popular circles. Essentially the problem is caused by the release of enormous amounts of nitrogen and sulphur oxides into the environment. These compounds have the effect of making water much more acidic, and indeed rainfall with acidity as high as 2.4 has been recorded in Scotland. This amount of acidification in rainfall has as one of its effects the destruction of surface vegetation through interruption of the metabolism of photosynthesising species – i.e. trees and plants die. Of course things are not as simple as this: the natural alkalinity of local environments can counteract elevated acidity in precipitation. Thus susceptibility to acidification is variable over the earth's surface. Nevertheless, even governments have recognised the impacts of SO_2 and NO_2 emissions on aquatic environments, and imposed strict limits on maximum allowable concentrations and developed programmes for public education and local mitigation. Studies in Canada and the UK have demonstrated that increased acidification is strongly correlated with reduced salmonid returns to spawning streams and reduced reproductive success (McDonald and Kay, 1988).

In Europe there is considerable concern about the effects of intensive acidification of the so-called "Black Triangle" region formed at the borders of the Czech Republic, Poland and the former German Democratic Republic (East Germany). The coal-fired boilers of public and industrial power generation, cogeneration, and district heating plants were, during the communist period, the largest sources of pollution in Central Europe. Ten power plants in this heavily-industrialised, 12,356 square mile region, producing 14,860 MWe and 4,000 MWt of energy; yearly burn 80 million tons of lignite coal; making a major contribution to the 3 million tons of SO_2 and 1 million tons of NO_2 emitted annually in the region. A study released in 1995 by the firm EEA Corinair showed that Poland, the Czech Republic and the former East Germany released nearly 9.5 million tons of SO_2 and more than 2.7 million tons of NO_2 in 1990 – before any real progress of post-Cold War environmental cleanup efforts could be made.

Microbiological Pollution

Microbiological pollution comes in a wide variety of forms, ranging from sewage pollution, to the creation of water habitats that permit the overabundance of certain plant or animal communities. Like physical and chemical quality assessment, there are a small number of key indicators used by water management authorities to assess microbiological WQ. They are discussed briefly below and in Appendices 2 and 3. Methods for collecting and analyzing this data were introduced earlier in this chapter.

Biochemical oxygen demand, abbreviated as BOD, is a test for measuring the amount of biodegradable organic material present in a sample of water. The results are expressed in terms of the mg/L of DO which microorganisms, principally bacteria, will consume while degrading these materials. For reasons discussed earlier, the depletion of oxygen in receiving waters has historically been regarded as one of the most important negative effects of water pollution. Preventing these substances from being discharged into our waterways is a key purpose of wastewater treatment. However, because the test takes too long to be useful for short-term control of the plant, the chemical or instrumental surrogate tests are often used as guides. Unpolluted water $=< 2$ mg/l 0_2; raw sewage ~ 600 mg/l 0_2; treated sewage effluent $\sim 20\text{-}100$ mg/l 0_2.

Another excellent indicator of water quality is based on the number of coliform bacteria. Coliform bacteria normally live in the intestines of mammals and are excreted with fecal wastes. Some forms are pathogenic but even if they are not, if they are present in a sample this indicates that the sample has been through an animal intestine. Wastewater treatment plants are often required to test their effluents for the group known as "fecal coliforms," which include the species E. coli, indicative of contamination by material from the intestines of warm-blooded animals. The US EPA does not allow any coliform bacteria in drinking water and EC standards have recently been revised to match.

Still another way of measuring the biochemical health of water is to assess the total amount of dissolved oxygen that it contains. The way that the oxygen levels fall as a result of an increase in organic material can best be shown by Figure 4.3, (Oxygen Sag Curve). The curve indicates when or where organic material was deposited (e.g. by a sewage works), its gradual elimination by micro-organisms and invertebrates and the natural purification by the water. Near the point of discharge the oxygen will be used up by the micro-organisms. Downstream a natural purification takes place as the effluent is diluted. This is known as *natural repair* and is related to the idea that natural watercourses have a certain capacity for self-purification discussed earlier. Areas of water such as ponds or lakes, in which the water does not move, suffer from pollution more than rivers as they are not able to get rid of the pollutants as effectively as a river, or any other fast moving water body.

As the water becomes more polluted, there are more organic nutrients available for the bacteria to break down which leads to an increase in mineral salts. This is known as *eutrophication*. However, the breakdown of these nutrients leads to a depletion in oxygen and an increase in CO_2, which can lead to the death of many of the other organisms such as fish and invertebrates, which depend on the ecosystem remaining balanced. Eutrophication encourages the growth of algal blooms, which cover the surface and prevent light from reaching the submerged plants in the water thus preventing photosynthesis. The problem is greatly enhanced at night-time, when plants have been unable to produce sufficient levels of oxygen for the invertebrates to breath.

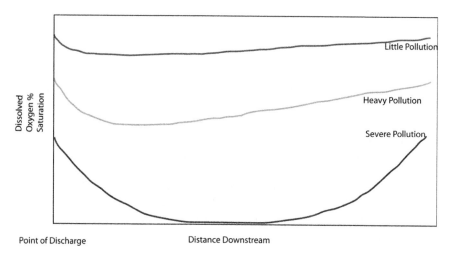

Figure 4.3   Oxygen sag curve

Eutrophication has the following effects on the ecosystem:

1. Affects animals, by causing lower oxygen levels, too low for animals to respire effectively,
2. Increases turbidity/cloudiness, reducing light levels and thus reducing photosynthesis by plants,
3. Increase substrate (fine muds), which block the gills of many organisms, especially benthic ones.

Several approaches can be taken to solve eutrophication. The symptoms can be treated, by removing macrophytic plants or by poisoning algal growths with chemicals. In a reservoir, mixing and aerating the water makes the algae easier to filter (because aeration causes bigger clumps to form, a process called flocculation).

Water and Wastewater Management in a River Basin

In many countries the quality of water resources, particularly surface water, is unsatisfactory. As a result it may be necessary to build clean water transport systems (sometimes over long distances) or to considerably increase expenditure on water treatment. In such cases there is an urgent need for clarifying the water-wastewater management within the river basin at the lowest possible cost. This means development and implementation of water quality improvement programmes (strategies) which set out the scope and schedule for the implementation of work geared to reduction of pollution loads discharged into the rivers from individual sources (location of WWTP and choice of technologies, etc.). Development of such investment programmes, particularly for larger rivers, requires the use of complex IT tools. There are numerous different computer models of surface water quality but the most popular and generally accessible one is the QUAL2K model (Brown and Barnwell, 1987). QUAL2K is an example of an open source ready-made software operating in a Windows environment, allowing simulation of the quality of river water for 15 water quality parameters including dissolved oxygen, biochemical oxygen demand, temperature, chlorophyll, organic nitrogen, ammonia, nitrate and nitrite, organic and dissolved phosphorus, bacteria from the E. Coli group and a few others.

Computer models such as QUAL2K optimise decision-making models providing answers to questions on what, where and when to build, to achieve the assumed effect at the lowest possible cost. We can distinguish two basic types of decision-making problems:

- *Minimisation* of the cost of wastewater treatment plant (WWTP) system within the catchment; Solution to this problem consists in defining the required level of wastewater treatment for each pollution source to ensure that pollution concentrations in river water at chosen monitoring cross-sections are lower than the predefined threshold values. In this task the decision-making variables are the wastewater treatment levels at each pollution source, objective function subject to minimisation defines the sum of total discounted capital expenditures and operating costs of all WWTP in the studied system, whereas the limitations describe the requirements as to water quality at chosen monitoring cross-sections.
- *Optimisation* of the schedule for the construction of a basin WWTP system; Solution to this problem consists in defining the dates (e.g. the year) of commencement of WWTP construction at specific sources of pollution so that the sum of total discounted capital expenditures and operating costs in the economic calculus period is the lowest possible, with the restriction that water quality in individual years must reach the assumed level of purity.

Though the QUAL2K model was developed by the US EPA for use within the United States, it can be adapted for use in other countries by importing location-specific flow information. A very simple case study of water quality modelling in an English watercourse is briefly presented below, with a more complex, albeit hypothetical, exercise in water quality modelling reserved to Appendix 3.

The River Kennet runs 70 km from Marlborough to Reading in Wiltshire where it joins the river Thames (Booker, 2006). Water from the river is abstracted before being treated in a water treatment works and then sent out to people's homes as clean drinking water. As well as drinking water abstraction there are other pressures on the river Kennet, it is the main outflow for treated water from Reading sewage treatment works and much of the river is either surrounded by agriculture or towns that have built up around it, all of these factors can add to an increase in river

Table 4.3   Water quality in the River Kennett, Wiltshire, England, December 2006

Pollutant	Maximum level for drinking water	River sample
Pesticide	0.1μl (individual) 0.5μl (total)	0.0805μl
Nitrate	50mg/l	7.2mg/l
Ammonia	0.5mg/l	0.04mg/l
Turbidity	10ppm	15.0ppm
pH	6.5-8.5	7.86
Temperature	25C	11.04C

pollution. Rainfall in the region is the nearly the lowest in the country and the river Kennet receives most of its recharge water from groundwater sources.

Sampling of water from the Kennet took place at a location just upstream from the WWTP at Fobney in December 2006 and revealed little of concern other than elevated turbidity levels – which could well have been a result of recent heavy rains. Nitrate levels were low, possibly also reflecting the time of year (fertiliser applications generally take place between May and July). Nevertheless the Kennett is considered by the Environment Agency to be at risk of both point and non-point source pollution primarily because of population density.

Concluding Comments

The role of water management consists in initiation and stimulation of development of water quality improvement strategies and programmes. Analyses carried out during development of such strategies should allow (through hierarchisation of water quality protection actions) for the most effective use of the available means. Additional tasks for protection of water quality faced by water management include primarily: development of international cooperation in respect of monitoring and preventing pollution of boundary and trans-boundary waters, ensuring efficient operation of quality monitoring systems for water and wastewater discharges, monitoring and stimulation of activities within the areas of underground water feeding, supporting and development of low-waste, water-effective and clean technologies, recycling and technologies for wastewater and sludge treatment and utilisation, supporting and development of educational programmes in respect of reduction of specific water use and the household use of agents which are less harmful to aquatic environment as well as the use of chemical agents in agriculture.

Suggested Activities for Further Learning:

1. Research some of the range of water quality standards around the world. The EC, World Health Organisation (WHO), US Environmental Protection Agency (USEPA) and the Canadian Department of Health guidelines are all readily available on the web, and many water management jurisdictions also publish their WQ data on their websites – see for example the English Environment Agency's on-line database of English surface water quality.

2. Campaigning organisations like Friends of the Earth and the US-based Coalition against Household Toxic Wastes publish data about specific harmful releases or polluting industries. Look at some of the case studies on a website like FoE's Chemical Release Inventory (http://www.foe.co.uk/campaigns/safer_chemicals/resource/factory_watch/).

3. Don't forget the more complex, mathematical exercises in Appendix 3, not included directly here as they are a bit lengthy! The Streeter-Phelps formula for degradation of pollution in a watercourse defines local water quality as a function of distance from the last point source and the speed of the watercourse.

4. Visit "Flush Gordon's Dirty Water website" (http://www.geocities.com/RainForest/5161/) for more about design of modern WWTPs. Consider the range of treatment processes, their relative costs and efficiencies, etc. You will also find there links to other websites dedicated to similar issues – perhaps it would be worthwhile to visit some of them?

5. Visit website of The Water Quality branch of Environment Canada (http://www.ec.gc.ca/water/en/manage/qual/e_qual.htm) which provides more general discussion of issues relating to water quality protection. Think over the following questions:

   – What are the consequences of poor water quality?
   – Eutrophication – what is it and when does it become a problem?
   – What are the means, ways and actions for protection of water quality?

6. Water quality testing to regulatory standards is complex business. However it is possible for non-specialists to get a good idea of the relative presence of most of the chemical compounds, water characteristics and biological indicators listed in the EU Water Quality Standards above. In this exercise we focus on techniques for the general characterisation and chemical analysis of water samples.

For this exercise you will need 10 sealable wide mouth water bottles, a "Palintest" kit (using "colorometer" comparison charts), and reagents for at least the following parameters: pH, sulphates, nitrates, cadmium and mercury. A turbidity comparison chart and thermometer will also be required.

Steps:

1. Reliable water quality testing of even a single water body such as a stream or river requires multiple samples taken over several points in time. If the sampling strategy is well-designed it should be possible to make inferences about relatively clean versus relatively dirty parts of the watercourse, diurnal variations, possible correlations with predictable peak time events, etc. Students should design a water testing regime for a small water body (lake/pond or stream/creek) near where they live. In collecting the samples care should be taken to use exactly the same process each time, filling the bottles from water undisturbed by footsteps, etc. and from the middle of the depth range. Label these samples with location, time and date and include any other data that might be relevant to interpreting the results (e.g. ambient temperature extremes of hot or cold).

2. Follow the instructions with the Palintest kit for testing for the listed substances and enter these values on a matrix. Also enter values for turbidity, colour and temperature using the standard classifications.

3. Now attempt to describe and explain the results. Are there areal variations? Diurnal variations? If so why? Are there perhaps correlations with peak time releases from industrial plants, or seasonal surges in nitrate applications in rural agricultural lands?

Appendices 2 and 3 offer more information and support for learning about water quality testing.

Chapter 5

Managing Scarce Water Resources: The Quantitative Dimension

Introduction

This chapter explores changing approaches to managing water in its quantitative aspect, that is to say we address the problem of planning for optimal allocation of increasingly scarce water resources. With ever-increasing water demand amongst an ever-increasing diversity of uses, and in a world of ever-increasing water scarcity, many countries in Europe and the world have reached the point of needing nothing less than revolutionary thinking. In other words, if we are to meet the quantitative challenges of water management in the 21st century we must first unthink the central ideologies and paradigms that led us to this impasse in the first place and then begin to construct their replacement. My argument is that the first evolutionary and then systematic approaches to water management which have dominated the past two centuries have been themselves dominated by what I call the "engineering paradigm" – the central preconception that any given problem can be solved with more and better technology. Thus we have moved from living in domestic premises which relied on perhaps 10-20 litres of water per person per day in 1900 to households that that require 140-150 litres/p/day in 2008. Of course part of this process has involved integrating better personal and public hygiene into our ways of living (Shove, 2003), but similarly we have developed a "natural" demand for more machines doing more things with more water along the way (Shove et al., 2007). And an ironic result may be that we are now becoming *less* healthy as a society. The same engineering paradigm that has populated our homes with all manner of washing and cleaning machines has also risen to the challenge of providing the increased volumes of water necessary to serve them. More demand means more supply, that is, reservoirs, treatment plants, pipelines, computerised control systems, etc.

In the next section I outline what I take to be the key features of the "engineering paradigm", utilising a broad range of examples from North America, Australasia and Europe. Subsequently the case of the Tennessee Valley Authority in the US, a classic example of high modernism in water supply management, is discussed in order to throw further light on the enduring power of the engineering paradigm. Further, the chapter presents a simplified standard methodology for solving problems of water management in a systemic manner. The idea of creating "regional water balance statements" is introduced a way of showing how difficult it can be to come up with optimal solutions to allocation of increasingly scarce resources. Finally, we turn our attention to the latest fad in water allocation: demand management. Here I shall argue that the current obsession (in the UK) with point of use metering diverts us from the real revolutionary potential inherent in demand management – it is in fact nothing more or less than a hijacking of demand-side management by the engineering paradigm.

As in previous chapters, throughout the text there are suggestions for further specific supporting reading. At the end of the chapter there are specific "activities for learning" which might be helpful in orientating your study process. Numerous quotations, tables, maps and illustrations are also included.

The Engineering Paradigm and Water Management

Humans have long thought that they could engineer themselves out of virtually any difficulty, and perhaps nowhere has that ill-fated logic been more manifest than in the area of water management. Ours are, at a fundamental level "hydraulic" societies; societies constituted largely out of the social, economic, cultural and political dimensions of control over water. Though invocations of Wittfogel are nowadays unfashionable in this respect he was at least empirically accurate: "Those who control the (hydraulic) network are uniquely prepared to wield supreme power."[1] Wittfogel's essential point was that the existence of a hydraulic network almost presupposed a high degree of centralised economic and political power and a cultural outlook that conceives of the earth as merely the raw materials for modernising societies. In this light comments (also noted in Chapter 1) such as those of Floyd Dominy, (former head of the US Bureau of Reclamation) "The unregulated Colorado was a son of a bitch. It wasn't any good. It was either in flood of in trickle." Camille Dagenais (Canadian dam engineer) "In my view nature is awful and what we do is cure it" and Paco Rabana (Chief architect of the Spanish Water Plan) "to let any of our rivers' precious waters reach the sea is a waste" are very much in tune with the modernist water ethos.

In the UK the "golden age" of hydro-engineering began in the last quarter of the 19th century with a succession of Acts of Parliament authorising the construction the Elan and Vyrnwy Valley reservoirs. By the end of the century numerous Welsh villages had disappeared, submerged by the waters rising behind new dam structures, and much of mid and north Wales was effectively transformed into a giant water reservoir for the rapidly industrialising English Midlands. The Vyrnwy Reservoir, completed in 1888, was the first major milestone, not just for its garnishment of Welsh water, but also because it was the first major engineered stone block dam; a technique of construction more material and cost effective and, with the inevitable neo-Gothic crenellations

---

1   Wittfogel, K., 1957 *Oriental Despotism: A Comparative Study of Total Power.*

added, a potent symbol of Victorian England's domination of Wales and its water (see Photo 1.2). Interestingly, ideals of purity and progress were linked to the waters thus brought to Liverpool:

> Now, happily, the crystal of the spring is brought to our doors, and vast lakes water sheds and rivers are "tapped", and by pipe lines, reservoirs and tunnels are made to supply water to our [Liverpool's] populations with is what is so needful for our welfare –water, plentiful and wholesome.

How ironic then that the water had to be "treated" with limestone to make it palatable to the good residents of Merseyside, who were accustomed to "harder" water (Hassan, 1998).

The (literally) crowning moment probably came when, on 21 July 1904 King Edward VII turned a ceremonial golden handwheel on the site of Elan Valley Dam, symbolically as well as literally starting the flow of impounded water eastwards towards Birmingham. The King was probably unaware that the dam he celebrated as "a modern wonder" had displaced more than 100 people, none of whom were compensated in any way by the dam builders. Similarly Vyrnwy, Arden and other large reservoir projects submerged Welsh villages and farms, with only rarely any compensation paid because these lands were mostly contained within large estates controlled by English land barons. No wonder then that Elan, Vyrnwy and similar schemes became potent symbols for Welsh nationalist politics, born about the same time.

More than the structures themselves however (and about which we will have more to say in Chapter 7), it is important to recognise that the ethos of what Erik Swyngedouw (2004) has called "hydro-modernity" rested on a deeper and more complex set of relationships. When the Birmingham Waterworks Corporation applied to Parliament for a patent to build the Elan scheme they were part of much more than merely slaking the thirst of burgeoning Birmingham. Elan Valley water, and by extension all large water developments of the era, were part of a wholesale reconstitution of urban life; from a model of cities as overgrown villages to new ideal of cities as protean entities greater than the mere sum of their inhabitants or the number of their factories. The new urbanism was bound up with the wholesale transformation of that de la Blache called "gens de vie". Thus, when Elan Valley water arrived in Victorian Birmingham it brought new ideas and ideals, about new ways to organise the household, new ways to relate to the environment, and new ways to exploit water for commercial gain. There is no coincidence in the arrival, simultaneously with King Edward's turning of the golden handwheel, of new rules for using water in the home as well as mechanisms for spreading these doctrines and enforcing them (if need be). Newly established public health authorities, National Schools and civil society organisations such as the Women's Institute inserted "public water" into Victorian England's homes in new and unprecedented ways – after all if the solution to the cholera epidemics of the preceding generation and the Great

Stinks was public water, then surely more water would lead to even healthier living. As Shove (2003), Hassan (1998) and others have shown, new norms of bodily health and domestic sanitation shaped increasingly water-intensive technologies such as toilets, baths and showers and, just outside the front door, even the garden. Somewhat ironically, a century later a new ethos, this time of water shortage, is driving moves towards efficiencies in water usage, such as new public health information about the benefits of fewer baths and showers, more water efficient clothes and dish washing machines and garden planting more attuned to local ambient water availabilities. We will come back to some of these issues later in this chapter and again in the final chapter of this volume.

If it can be argued that the movement of large volumes of water from the country to the city marked a new urban – as well as a new hydro – modernity, then it is also true that this moment was constituted through what David Harvey aptly calls the "urbanisation of capital" – the permeation of commodity capitalism into our most intimate spaces and places. Water, after all, became necessary in larger and larger quantities at exactly the same moment that it became an industrial, a social and a cultural commodity. Was it merely by chance that, for example, that new water-intensive household appliances came onto the market at the virtually the same moment that industrial water reached into the hearts of modern cities? Was it merely happenstance that within a generation the necessity of these appliances would become encoded in, and indeed prescribed by, urban building regulations? And does it not seem exceptionally "tidy" that the state should be the one to take on the initially risky and tremendously expensive task of creating the water infrastructure that many contemporary experts ardently wish to "privatise"? In the next chapter we will explore the terrain of water privatisation, in Europe and around the world, but for present purposes it is important to consider more deeply the historical geography of water's state-sponsored industrialisation and commodification ... and its implications for the challenges now facing us in the 21st century. And a good place to start is with a detailed look at one of the boldest and best documented instances of state-sponsored hydro-industrialisation ever – the Tennessee Valley Authority of the eastern US.

Case Study: The Tennessee Valley Authority

The Tennessee Valley Authority is one of the largest and most well-known examples of comprehensive integrated water resource management. Some critics see it as essentially a hydropower mega-corporation, and it has indeed constructed several dozen dams since its inception by US Congress in 1933. But its remit is much broader, including also navigation, drinking water supply, irrigation, and regional economic development. Its history is instructive.

During the 1930s, US President Franklin Roosevelt needed innovative solutions if the New Deal was to lift the nation out of the depths of the Great Depression. And the TVA was one of his most innovative ideas. In the 1930s, the South was a poverty-stricken section of America that had escaped the wealth accumulated in other regions of the country.

In 1933, for example, the per capita income in the Tennessee Valley was only 45% of the national average. Thousands of rural southern people had no electricity, or running water, suffered failing crops year after year from depleted and eroding soils and lived in abject poverty. Additionally, the large timber companies had cut the best timber using the common practices of "high-grading" and clear-cutting (McCaleb, 1999). The nation as a whole was reeling from the devastating impacts of the Great Depression, and the South, particularly the rural South, was in a desperate struggle for survival.

Roosevelt envisioned TVA as a totally different kind of agency. He asked Congress to create "a corporation clothed with the power of government but possessed of the flexibility and initiative of a private enterprise". On 18 May 1933, Congress passed the *TVA Act*. It had five key purposes:

1. The maximum amount of flood control;
2. The maximum development of the Tennessee River for navigation purposes;
3. The maximum generation of electric power consistent with flood control and navigation;
4. The proper use of marginal lands;
5. The proper method of reforestation of all lands in said drainage basin suitable for reforestation; and
6. The economic and social well-being of the people living in the Tennessee River basin.

Right from the start, TVA established a sort of missionary zeal to fulfilling its mission-development through resource exploitation and engineering. TVA developed fertilisers, taught farmers how to improve crop yields, and helped replant forests, control forest fires, and improve habitat for wildlife and fish. The most dramatic change in Valley life came from the electricity generated by the rapidly growing network of TVA dams. Electric lights and modern appliances made life easier and farms more productive. Electricity also drew industries into the region, providing desperately needed jobs.

During World War II, the United States needed aluminium to build bombs and airplanes, and aluminium plants required electricity. To provide power for such critical war industries, TVA engaged in one of the largest hydropower construction programs ever undertaken in the United States. Early in 1942, when the effort reached its peak, 12 hydroelectric projects and a steam plant were under construction at the same time, and design and construction employment reached a total of 28,000. By the end of the war, TVA had completed a 650-mile (1,050-kilometer) navigation channel the length of the Tennessee River and had become the nation's largest electricity supplier.

Significant changes occurred in the economy of the Tennessee Valley and the nation, prompted by an international oil embargo in 1973 and accelerating fuel costs later in the decade. The average cost of electricity in the Tennessee Valley increased fivefold from the early 1970s to the early 1980s. With energy demand dropping and construction costs rising, TVA cancelled several nuclear plants, as did other utilities around the nation.

In 1998 TVA unveiled a new clean-air strategy to reduce the pollutants that cause ozone and smog. It has also begun to sell leases and rights to develop recreational and other activities around its reservoirs, such as the 1997 deal to develop a Golf, Leisure and private residential complex along the shore of the Nickajack Reservoir. The initial plans called for 500 homes and cabins, 120-room resort lodge, an 18-hole golf course and a 300-slip marina.

Today, TVA manages 480,000 acres of public land including Land Between the Lakes (LBL). Of those acres, 80% is allocated to recreation and resource conservation. TVA operates a total of 54 dams and reservoirs including 35 multipurpose dams on both the Tennessee River and its numerous tributaries. A full 52% of 480,000 acres is located along 11,000 miles of shoreline. TVA estimates that visitors each year to TVA reservoirs and marginal lands contribute $1.25 billion to the Valley's economy. In the 1990s the TVA also showed real interest in the development of wind farms, starting with the 2 MW capacity recently built at Buffalo Mountain, TN.

A Systems Approach to Water Resource Management in the 21st Century

Water is generally considered to be a renewable resource. This is of course true in a regional sense but in a global perspective water resources are confined to a closed system. On a global scale there are no water incomes, except possibly the chemically bound water contained within meteorites (!), and there are no water expenses, either, except for the occasional water molecule entering free space. According to Peter Gleick (1996) the global water cycle involves approximately 1.4 billion cubic km of total water, of which 50 million cubic km is freshwater or about 3% of the total. Of this 50,000,000 cubic kms only about 100,000 cubic kms are available as surface or groundwater sources. As a rule of thumb we use 150-200 litres/capita/day in households and twice as much in industrial and agricultural applications. The challenge is to meet the need with basic renewable resource which is unevenly distributed geographically as well as temporally.

There is a danger in viewing water resources from an excessively narrow regional perspective. Since water is circulated globally in the hydrological cycle, all the water we use has a geography as well as a history. If we pollute the water regionally, for example, the effects may show up in a totally different region with unforeseeable environmental effects. Pesticides used intentionally in farming in one region may cause the death of birds elsewhere, e.g. in the Baltic Sea. Another example is when sulphuric smoke pollution and the humidity in the air in coal-burning industrial regions kill the coniferous trees in distant forest regions, thus altering the hydrological balance locally. On a larger scale, studies of the El Nino Southern Oscillation have shown that minute variations in water layer temperatures in the South Pacific Ocean can dramatically alter precipitation patterns around the world – leading to droughts in some places and floods in others (this issue is taken up in more detail in Chapter 8). These causes and effects are well known today but were either not foreseen or were ignored at the time when contemporary water management practice was initiated.

The systems approach can also be taken when studying the main users of water resources: agriculture, industry, and urban areas.

In agriculture, the water incomes are essentially defined by (largely natural) precipitation and (largely human) irrigation. The expenses are evapotranspiration, drainage and groundwater formation. The strategy of water management is therefore twofold: to deliver an appropriate amount of water to the growing plants and to distribute nutrients to the plants' roots using water as the transport medium. If there is a problem of water shortage, irrigation is applied using groundwater or a nearby river, while if there is too much water the field is drained. The principal action in both these respects is to divert the natural flow paths of the water. The second action, increasing the nutrient status of the field, adds another dimension to water management, i.e. water quality. High nutrient content means high quality for plants but low quality for humans, animals and the environment generally (recall the discussion of eutrophication in the preceding chapter). Further issues are erosion and salinisation but these problems are relatively minor in most of continental Europe. The environmental issues in this system can thus be defined as minimising the harmful effects of the diversion of the natural flow paths; and keeping the nutrient-rich water "inside" the system, where it constitutes a resource, rather than allowing it to escape to where it constitutes a risk.

In industry, water incomes and expenses are defined by the system of pipes entering and leaving the plant. The objective means using water either as a non-consumptive part, e.g. transport or cooling medium, or as an ingredient in the industrial process. The main problem is the addition of dissolved substances to the water, caused by the production process, even if it is only heat energy, which is carefully regulated in most European nations through abstraction licencing and monitoring processes. If the water is exported back to the same body of water from which it was imported, no significant diversion from its natural pathway has been made. However, used groundwater or surface water is normally redistributed via the municipal sewage treatment plant into a surface water body, often a different one than the one where the water originated. The main environmental issues are thus to minimise the impact of the industrial processes on the water and to minimise the water volumes handled. In recent years the Environment Agency for England and Wales has undertaken a "review of sustainable abstraction" precisely in order to reassess abstraction volumes against a changing hydrological regime and better scientific knowledge about the needs of the environment (Environment Agency, 2008). The optimal solution in some cases may be to close the system totally, avoiding strain on water resources and the export of polluted water. In the food industry, where much water is used for cleaning and for cooling, largely closed systems are not uncommon.

Urbanised areas, i.e. cities and towns, have properties that constitute a mixture of agricultural and industrial systems. The water incomes are precipitation, discharge and water from local or municipal wells or waterworks. Expenses are evapotranspiration, sewage and stormflow from sewage treatment plants or individual houses, water lost in the distribution system, locally infiltrated precipitation and losses of groundwater. Note that the system can be defined in different ways, depending on the problem studied. If the problem is the damage to buildings caused by the lowered groundwater level, then locally infiltrated stormwater constitutes a systems income and a potential solution to the specific problem. Several objectives can be identified. To deliver high quality drinking water and to treat sewage water are the main objectives, but to get stormwater off the streets, to avoid inundation by regulating surface water levels and in most cities preserving the groundwater level are also important objectives. Urban water planning and management is complex, involving the administration of both waterworks and sewage treatment plants as well as city and road planning. This area is of special importance since it interfaces directly with citizens and their need of household water for everyday activities. The environmental issues involved are those of supplying domestic water of sufficient quality and quantity, sanitation and the proper treatment of sewage water, handling of storm flows and prevention of inundation and lowering of the groundwater table.

Intensive growth in industrial production, the development of increasingly large urban/industrial agglomerations and intensification of farming often lead to such increases in water demand and water pollution that deficiencies of usable water become a barrier to further economic growth. Growing difficulties with the right solution of problems with the *supply of water* of suitable quality, complex links between various ways of water resource use and availability of these resources require comprehensive planning of the development of water management and the use of water resources within the framework of the so-called *water-economic systems*. A water-economic system may be defined as a functional/spatial system comprising the natural surface and underground water resources, the natural environment in which such water occurs and the *hydrotechnical structures* providing for developing these resources, water user facilities and links between these elements. Such a water-economic system is focused on meeting the *users' water demand*, meeting certain environmental requirements and protection of certain areas against *flood*.

Water circulation within a water-economic system is controlled by hydrotechnical facilities (storage reservoirs, transfer channels, wastewater treatment plants) which fulfil such functions as transformation of water resources in time and space and modelling the quality of those resources. Because of the high capital expenditures and a relatively long time of depreciation of hydrotechnical facilities it is necessary that the use of these facilities in water resource management is as comprehensive (rational, optimum) as possible.

Regardless of their individual specificity resulting e.g. from the climate, land configuration or the degree of urbanisation each water-economic system, in terms of management theory is a complex system, i. e. it is a complex of numerous interlinked facilities such as rivers, protected areas, lakes and storage reservoirs, rural settlements, urban agglomerations, industrial plants, agricultural areas, etc.

In terms of water management potential the following features are the most important ones in water-economic systems:

- The high degree of complexity expressed by a large number of facilities the majority of which are complex systems in themselves (e.g. the water supply system in an urban agglomeration including a water intake, water treatment plant, compensation reservoirs, water supply network, pumping stations, etc.);
- Presence of facilities with highly diversified dynamics (e.g. large reservoirs with dynamics measured in monthly scale, pumping stations or compensating reservoirs with dynamics measured on an hourly scale, etc.);
- Physical distribution of facilities over large areas which is the reason why links between certain facilities or groups of facilities may be considered weak or impossible (not least because of attendant energy costs[2]);
- Qualitatively different nature of tasks of the system and operational rules of hydrotechnical facilities in regular and emergency conditions, such as floods, system failures resulting from damage in its facilities (dams, wastewater treatment plants, etc.);
- The magnitude and diversity of its objectives (water supply, recreation, water quality protection, production of electric energy, etc.);
- The increasingly unpredictable nature of the water inflow into the system and water demand of certain types of water users (e.g. irrigation in agriculture depends on the current meteorological conditions).

In view of the above definition of the water-economic system, its properties and the objectives it has to achieve it should be noted that planning and controlling such systems require the application of system analysis techniques. For this purpose it is convenient to split the relevant water resource management issues into four interrelated key groups:

1. Water Resources;
2. Water Management Tasks;
3. Measures (Water Management Tools);
4. Criteria for the Assessment of Actions taken.

---

2   Some water company executives in England worry that pressure to reduce the carbon footprint of their operations will impact disproportionately on those operations located in regions requiring significant pumping – such as the English Midlands and Black Country.

Water Resources

Water resources include the natural underground and surface water resources and the so-called production waste water (e.g. mining water). Water resources are variable in time and space. Variability in space relates both to surface and underground water resources. In respect of variability in time it should be noted that the dynamics of changes in surface water are more volatile than that of underground water. Surface water resources (river runoff) depends on the volume of precipitation and evaporation which are featured by very significant variability in time. Resources cannot be completely used up for economic purposes for two reasons: the necessity of preserving the so-called *base flow* in water courses and high flow variability in time (in some rivers the maximum flows are more than 1,000 times greater than the minimum flow values). Figure 5.1 shows annual variation across only one year, 2006, for the River Kennet as measured at Theale, near Reading (this is the watercourse for which quality data was presented in the last chapter).

For these reasons the term *"available water resources"* has been introduced: it describes that part of resources which may be drawn from rivers on a permanent basis, observing a certain level of *guarantee*. The notion of available resources has also been introduced in the determination of underground water resources and it means the part of resources (corresponding to natural feeding) which may be drawn without disturbing the dynamic equilibrium. Exceeding this limit may lead to a drop in the underground water table and to their complete depletion.

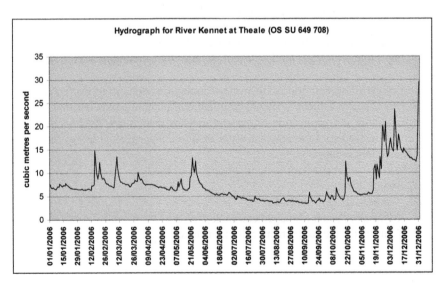

Figure 5.1   Hydrograph for the River Kennet at Theale, England, 2006

Source: Based on Environment Agency data.

The natural processes that affect the water resources are described and modelled by *hydrologists* who are, as we shall see in Chapter 8, increasingly concerned about the effects of climate change.

Water Management Tasks

In terms of the quantitative dimension the main management tasks in a given area (catchment, region, country) may be split into:

- Protection against water-related risks in the form of floods (from rainfall, meltwater, storm, ice jam, caused by hydrotechnical facility failure or maloperation) or drought;
- Protection of water resources against pollution and excessive use, excessive modification of hydrological regime and the loss of natural values of water and nearby areas as well as concern for preservation of balance in the natural circulation of water in nature;
- Meeting the *water demand* of various types of water users including population centres, industrial and agricultural users, transportation, recreation and the environment.

Among the above specified economic uses of water resources we may distinguish two basic types: *consumptive uses* which partly or totally use up the drawn water (industry, municipal management, agriculture) and *non-consumption uses* which do not affect the volume of water resources (water power engineering, inland navigation, recreation, tourism and water sports).

As we have seen the preceding chapters, management for different objectives or functions is, in England and Wales, undertaken by a functionally differentiated state apparatus. Thus whilst the Environment Agency is responsible for overall water quality management and qualitative abstraction, the Drinking Water Inspectorate pays particular attention to drinking water supplies, British Waterways manages the extensive canal network and Natural England works to protect water as part of the landscape.

Measures (Water Management Tools)

Water management instruments may be split into *three* key groups:

- **Technical measures:** (hydrotechnical structures) which ensure developing of the distribution in time and space of available resources and their quality. Among these structures the following are the most important:
  - *storage reservoirs* which allow storage of water in high water periods and making it available to users during the periods of dry weather flows in rivers (thus the reservoirs perform two fundamental tasks: elimination or mitigation of the flood risk and improve the water supply conditions);

- *transfer channels* allowing transfer of water from regions with abundance of water to areas with water deficits;
- *wastewater treatment plants* (mechanical, biological, chemical) which ensure reduction of sewage-contained pollution discharged into surface water.

- **Legal and administrative measures:** legal standards determine the scope of tasks, competence, responsibility and duties of governmental and self-governmental authorities and administration bodies. These standards also define the rights and responsibilities of water users. The rank and scope of individual standards depends on the organisational structure of integrated water resource management adopted in a given country. The following are among the most important standards:
  - standards regulating the principles of water resource use;
  - standards regulating the principles of environmental use;
  - standards regulating the physical planning issues.

For instance, in Poland the most important standards include the following:
- a system of legal standards regulating the principles of water resource use ("Water Law" Act of 1974);
- system of legal standards regulating the principles of environmental use (the 1980 Law on environmental protection and management and the 1991 Law on the Official Environmental Protection Inspection);
- system of legal standards regulating the physical planning issues (the 1994 Law on spatial management).

In Chapter 3 we examined the equivalent UK and EU structures.

Legal standards of the highest rank regulate the principles of water resource use and define i.e.:
- the types of water property, the rights and responsibilities of its owners;
- water management bodies;
- the principles of water quality protection (standards for wastewater discharged into water or soil, quality requirements for water for various uses, underground and surface water intake protections zones);
- underground and surface water quality monitoring and classification principles;
- the principles of flood control;
- the key instruments for water resource management (*river basin* management plan, the conditions for water use in the basin, *water permits*, a system of charges and fines connected with water resource use, water management IT system – water management register).

European Union Directives are examples of standards regulating the principles of water resource management.

The measures of an administrative nature include:
- standards on water protection (e.g. the level of *base flows, harmless flows*);
- flood control arrangements (*flood hazard zones*, bans and restrictions on management of flood areas, *flood reserve capacity* in storage reservoirs);
- principles of operating hydrotechnical facilities (e.g. water management instructions for storage reservoirs).

- **Economic measures:** are focused essentially on stimulation of user activities along lines corresponding with the national policy for water resource use and management. The role of these instruments in the water resource management policy is essential and steadily growing. They efficiently complement the legal and administrative regulations discussed above. The most frequently used economic measures include the following:
  - charges for water intake;
  - charges for wastewater disposal;
  - charges for the use of water facilities;
  - fines for non-compliance with requirements in respect of wastewater disposal;
  - subsidies granted for project implementation in the scope of water management and water protection;
  - tax and tariff diversification;
  - product-specific charges for products harmful to the aquatic environment;
  - trading wastewater disposal permits;
  - ecological insurance;
  - diversification of insurance contributions depending on the level of flood hazard.

Criteria for the Assessment of Actions Taken

Water management systems undergo continuous modifications due to: expansion of hydrotechnical structures, introduction of new users, modification of water management principles in existing hydrotechnical facilities, improvement of operational decision making processes. The associated alternative investment solutions and organisational and administrative changes must be compared, i.e. assessed in terms of various criteria allowing complete or at least partial segregation of feasible alternative solutions to facilitate the choice of the best approach.

The criteria applied in water management may be split into two key groups:

- Non-economic criteria, describing the level of implementation of tasks set forth for the discussed water-economic system. Examples of such criteria include: the level of meeting the users' water needs (e.g. guarantee), water quality in a chosen test profile of the river (e.g. dissolved oxygen, 5-day biochemical oxygen demand, pH, TSS, total phosphorus, total nitrogen), attractiveness of the water reservoir for recreation purposes (e.g. the range of water table fluctuations during the holiday season), modification in the river hydrological regime (e.g. changed value of characteristic flows, changed frequency of valley flooding as a result of possible implementation of a given hydrotechnical project). The above examples show that these criteria relate to quantitative and qualitative characteristics of the initial conditions and states of the discussed water-economic system;
- Economic criteria, defining in monetary units the effects (or results) of the given alternative of water resource management. Examples of this type of criteria: capital expenditures and operating costs of hydrotechnical facilities, the value of prevented flood losses, losses resulting from water deficits encountered by users, mean annual value of generated electric energy, etc.

The variety of effects and results of undertakings within the water management framework means that assessment of individual alternative solutions to the problem is often of a *multi-criterion* nature. This type of analysis may be exemplified by the choice of principles (method) of water management in a *storage reservoir* which should fulfil the following functions: ensuring the *base flow* in the river downstream of the reservoir, water supply for a large industrial plant, flood control for areas located downstream of the reservoir, electric power generation in a water power station, recreation and water sports. In such cases the set of criteria could have the following form:

- Frequency of occurrence of flows not lower than the base flow;
- Frequency of occurrence of water deficits (deficiencies) in the industrial plant;
- The degree (magnitude) of reduction of flood wave peak flow with a defined probability of occurrence;
- Mean annual production volume of electric energy in the water power station;
- Mean annual storage reservoir operating costs;
- The amplitude of water table fluctuations in the reservoir during the holiday season.

The set of criteria presented above is exemplary only and in a specific case it may be expanded, certain criteria may be omitted or replaced by other ones (e.g. water quality criteria have not been taken into consideration). It should be emphasised that the choice of the set of criteria is one of the most essential stages in the decision making process and may seriously affect the optimal solutions reached.

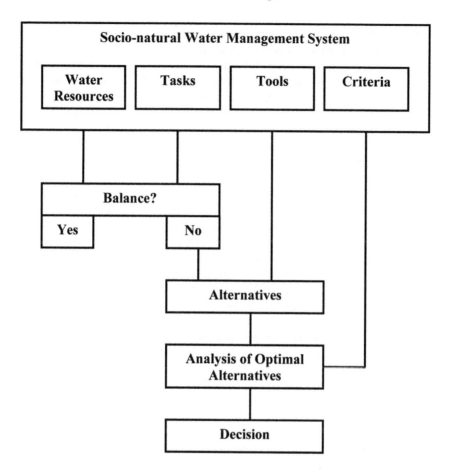

Figure 5.2　Holistic water management decision system
Source: Author.

Figure 5.2 shows the general scheme of procedures for solving water resource management problems within a predetermined area and is effected in the following stages:

- *Identification of the state of water management* including the definition of the scope of analysis and collection of input data describing the water resources and water management tasks complete with requirements in respect of the degree of their implementation (e.g. series of mean daily flows and the volume of the user's water needs which should be met with a certain *guarantee, hydrograms* of flood waves and the level of *admissible flow* in a given river section).

- *Identification of water management problems* including a comparative analysis (*water-economic balance*) of water resources and tasks (e.g. comparison of flow series with the volume of the user's water needs to compute the guarantee of meeting these needs and comparison with the requirements set forth by the user). This stage serves the purpose of recognising whether the studied system faces a problem consisting in the lack of balance between water resources and *water management tasks* (e.g. problem with ensuring water supply for the user with required *guarantee*). In case no such problem is identified the analysis is terminated – otherwise the next stage must be performed.
- *Definition of the set of alternative solutions of the problem* including development of various scenarios of balancing the water management tasks and water resources with the use of technical (e.g. construction of a storage reservoir, advancement of production technology towards water-effectiveness, construction of dikes, etc.) and legal/administrative measures (amendment of the instructions for water management in a storage reservoir, insurance system for areas of enhanced flood risk, requirements for discharged wastewater, etc.). Within the framework of this stage simulation-optimisation investigation will be carried out to define the technical parameters of proposed solutions (e. g. capacity of the proposed storage reservoir, the type and efficiency of the proposed wastewater treatment plant, etc.).
- *Choice of the best solution of the problem* including analysis of individual alternatives in terms of the set of the following criteria: economic (capital expenditures and operating costs), natural/ecological (required maintenance of base flows, modification of hydrological regime in the river, quality of river water, etc.) and social (number of new jobs, increased safety, etc.).
- *Decision and its implementation and monitoring of implemented project.*

Water Balance Statements

At various stages of the preceding chapter, readers might well have gotten the idea that there should be some sort of "balance" between water available from sources and water used from those sources. In Chapter 1 this idea was introduced through the observation that many nations around the world were increasingly finding themselves in positions of water "scarcity" or "shortage" as their needs outstripped their sources. And above it was implicit, I think, that there should be some sort of balance between the amount of water taken by a given user, or collection of users, and the available resources. Indeed, the situation presented in the last section becomes much more complex if we introduce more "real world" recognitions that there will be a wide variety of users, and that their needs, qualitative as well as quantitative, may be to some degree incompatible (e.g. waters that cannot be used for drinking water supply might be acceptable for irrigation or industrial uses – so-called "conditionally clean" water). Moreover, if we attempt to incorporate flood control, aesthetic, leisure and ecological "uses", and other sorts of supply such as more effective water recycling (or reduction of consumptive activity), the situation can become yet more complex. In practice water managers are faced with the ongoing challenge of optimising the use of limited water resources among competing uses and users. One tool for looking at issues of "balance" and how to achieve it is the "water balance statement", in particular as developed by Stephen Merritt and his colleagues in the Water Research Group at SOAS in London (Merritt, 1999, but see also Penkova and Shiklomanov, 1998). In this section we briefly present the logic of the water balance statement as a tool for water management at the level of regions or catchments (which will often be the same thing).

The water balance statement will generally relate to catchments (particularly with the advent of integrated water resource management, see page 137). The basic methodology involves compiling a spreadsheet with four columns: the first two are the categories of *supply* and the *quantity supplied per unit period of time*; the second two are the categories of *use* and the *quantity used per unit period of time*. For simplicity, in Table 5.1 and 5.2 there is only one user, a small town of 2,000 people, and only two supply sources: *ground water* and of *surface water*. From these are deducted the leakages and evaporation in the supply system, that is, between the points of abstraction and the supply/use boundary.

The calculation of flows in the regional water balance statement requires a form of "double-entry water accounting". The statement is treated as a single ledger containing all the appropriate entries either as supply items on the left-hand side or as use items on the right-hand side. For any flow of water to be recorded in the statement, it must qualify as some hydrosocial category of input to the regional system. In the water accounts ledger, with comprehensive and accurate records, *the statement should always balance*. In Tables 5.1 and 5.2 I present a greatly simplified example, based loosely on the hydrograph for The River Kennet presented as Figure 5.1 above. In Table 5.1 you can see the supply categories for a small catchment with few complications. The largest source of supply is, of course, the River Kennet itself and we are drawing half a megalitre per day from it. To this is added the 350,000 litres of groundwater we are pumping and the 100,000 we are recycling from water treated and returned to the Kennet. In this simplified model we are neither exporting water from our little catchment nor are we importing it. The only other thing we must account for is supply system leakage (and which provides some of the return flows), which is generally on the order of 15 to 20%. Taken together this gives us a total daily supply of 0.688 megalitres or 688,000 litres.

Table 5.2 shows the components of daily demand. In our hypothetical case there is only one populated place with a population of 2,000 people. Since we know that in England per capita use is somewhere between 140 and 150 litres/person/day we can estimate that the domestic demand will be approximately 300,000 litres. We have assumed relatively modest demand from other significant users, agriculture, commerce, public services and "other". The result is a daily demand of 650,000 litres.

Table 5.1   Supply categories

Categories of supply	ML/day
Ground Water Abstraction: 1st time through	0.35
Ground Water Abstraction: recycled sources	0
Surface Water Abstraction: 1st time through	0.453
Surface Water Abstraction: recycled sources	0.01
Import of Water from Another Catchment	0
Subtotal	0.813
Less: supply leakage and evaporation	0.1626
Less: export of water to another region	0
TOTAL NET SUPPLY	**0.65**

Table 5.2   Consumption categories

Categories of Use	ML/day
Households	0.3
Agriculture	0.05
Commercial Sectors	0.1
Public Services	0.1
Other Uses	0.1
TOTAL USE	**0.65**

Now, the "balance" part of the water balance methodology involves ensuring that the supply listed in Table 5.1 exactly match the demand listed in Table 5.2. Since they look like they do this seems like good news. There are however a number of problems. First, according to these tables, during the summer months we will be taking virtually all the flow from the Kennet, something which is environmentally unsound, risky and the Environment Agency would not allow us to do it anyway. Second, we have made no allowances for variations in demand across the year. It is a near certainty that water demand will rise in the summertime and this of course stands to exacerbate the water shortage inherent in these calculations. Third, the daily water supply and use figures actually need to be broken down into hourly figures since both the sources of supply and the loci of demand actually vary quite markedly on an hourly basis. Thus domestic demand is highest between 6.30 AM and 8.30 AM in the morning and again between 4.30 PM and 6.00 PM. In this case it is a certainty that in these periods the taps would run dry unless…there was some water storage built into the system. Groundwater abstraction is the obvious storage source in this example, but it is likely that the need to maintain pressure and suppress turbidity across even this small network would entail one or more small supply reservoirs.[3]

A conceptually related initiative was recently undertaken in central England. The Corby Water Cycle Study, initiated in 2005, sought to develop a water balance model for the planned community of Corby in Northamptonshire. The water cycle strategy mapped out growth and its associated water demands against the capability of existing infrastructure to deal with it. The strategy also set out what infrastructure would need to be in place, by when and what it would cost. These needs are set out in Figure 5.4 and show how projected growth in one of the nation's designated growth poles will entail additional infrastructural capacity as well as an altered geography of water infrastructure.

Demand-side Water Management: An Emerging Paradigm

Above I introduced the idea of demand-side management as one development scenario for managing scarce water resources. The other major paradigm, more closely aligned with the Engineering Paradigm discussed above and in Chapter 7, is supply-side management. Whilst supply-side solutions to water management were dominant in the 19th and 20th centuries, since the Rio Summit in 1992 the clear emphasis of government policymaking has shifted to managing demand; to trying to alter the structure of incentives, penalties, and the fabric of the built environment itself such that water users do more with less water. This paradigm shift is compounded by the clear social trend in the UK towards more smaller households, which risks greatly inflating aggregate demand for resources, including water. Recent reports from The National Housing and Planning Advice Unit (NHPAU) and the Barker Review suggest that a figure of 3 million new homes over the next ten years may not be enough to overcome the current housing crisis in the UK, due to the underestimates of population rise through immigration and longevity (Barker Report, 2004; ODPM, 2003). Another driver for this

---

3   Examples of water balance analyses using some of these techniques at different spatial scales:

- Souss Basin, Morocco (http://www.ce.utexas.edu/prof/maidment/gishydro/ africa/ex6af/ex6af.htm).
- Aral Sea Basin, Central Asia (http://www.fao.org/ag/AGL/AGLW/webpub/ aralsea/aralcour.htm).
- Digital Atlas of the World Water Balance, 1999 (http://www.crwr.utexas.edu/ gis/gishydro99/atlas/Atlas.htm#introduction).

Each of these examples was developed by a different institution for a different purpose, so the precise specification of the model varies somewhat. But study of them will give you a good idea of how water balance statements are used in actual practice.

paradigm shift is our growing scientific knowledge of the likely effects of climate change upon the UK and the Government's planned shift towards carbon neutral homes to reduce carbon emissions. The Stern Report in 2006, followed by the 4th IPCC report in late 2007 controversially predicted that average global mean temperatures will rise from between 2-3 degrees by 2050 and that a proactive approach in tackling the effects of carbon dioxide emissions is the most viable long-term strategy.

In the past water resource management has been focused upon a supply side fix which generally saw the construction of large infrastructure projects such as large dams to maintain supply. The concentration upon supply became the main economic strategy while efficiency gains through reduced consumption patterns were seen as the less "sexy" option. Yet the Environmental Agency for England and Wales now suggests that UK water companies ought not to invest in expensive new supply. The EA estimates that the introduction of water metering, low flow showers, low flush toilets and a variety of other water-saving technologies could save approximately 65 mega litres per day in the South East and London. Parliament has established a cross-party "Water Saving Group" and the private water companies have all written conservation and demand management into their 25 year "water resources plans", published in summer 2008 (Appleby, 2006, p. 40; ODPM, 2003).

So all appear agreed that demand management, particularly in terms of the water technologies fitted into new housing development, is the new key to sustainable water management. But how to achieve it? Over the course of the 1990s, new development standards, such as "Eco Homes", PPG 12 and the *Water Supply Regulations* (1999) provided various forms of regulatory encouragement. The 1999 *Water Supply Regulations Act*, for example did make it mandatory for low flush toilets to be fitted to all new developments with the incentive of a ECA (enhanced capital allowance) to encourage developers to take a more sustainable approach to water management (Rawlinson and Langdon, 2007).

The economic valuation of water is a key element in demand-wide management. This argument is based on the suggestion that as a low marginal value (but high fixed cost) good, there has been little financial incentive for water users (as opposed to developers) to economise (Brouwer and Pearce, 2005; Howe, 1971; Young, 2005). The 1992 Dublin Statement (discussed in Chapter 2) stated that "water is a scarce resource and has an economic value in all its competing uses". Such statements dovetailed nicely with the prevailing neoliberal ideology of the day and led not just to an renewed emphasis on water pricing mechanisms but to the wholesale privatisation of the water supply industry in England and Wales in the early 1990s (Howsan and Carter, 1996). Subsequently this "English model" of outright water privatisation within the context of a stronger state regulatory apparatus has been exported first to the postcommunist countries of Eastern Europe and then the rest of the world (this topic is discussed in greater detail in a subsequent chapter).

The results of all this sectoral restructuring and regulatory reform have been, in a nutshell, modest. Firstly, the proliferation of water-efficient technologies has been slow, with the government oddly (given its otherwise bullish statements) reluctant to ban water intensive technologies like high volume flush toilets (toilets consume about 25% of domestic water demand), low flow showers, aerators, etc. There has been some attention to rainwater harvesting and grey water reuse, but these technologies do not currently survive cost-benefit analysis as well as having specific technical problems. Grey water reuse systems, for example, raise water flow imbalance problems which can only be rectified by the installation of on-site storage tanks – easy at the point of initial construction, difficult to retro-fit – see Figure 5.3). However a study by Jefferson et al. (1999), showed that the greater the storage capacity, the greater the problem of bacterial formation and degradation within the storage tank.

The UK government has recently signalled that it is considering legislating for universal water metering in England, Scotland and Wales. Echoing statements from other government ministers and senior civil servants, the Climate Change Minister, who is drawing up the Government's water strategy, recently said: "The case for universal metering is now overwhelming – provided there is protection for low income and large families." This viewpoint is echoed by OFWAT and the Environment Agency in recent statements (OFWAT, 2005; King, 2006; 2007). Government officials contend that paying only for what you use (as measured by a meter) is both the fairest way to pay and it also offers the potential for significant conservation savings, an important issue as the balance between supply and population continues to shift in favour of the latter. New Labour, which argued vociferously against compulsory universal metering whilst in opposition, now believes meters should also help consumers spot leaks, as sudden rises in the amount of water used will be more obvious (Drury, 2007). The overall context for this renewed enthusiasm for water metering includes: the proliferation of larger numbers of smaller household units, the adoption of new more water intensive appliances (power showers, etc.), and the changing UK hydrological balance as a result of climate change (Hall, Hooper and Postle, 1988; Alcamo et al., 2007; Darrel Jenerette and Larsen, 2006; Mercer, Christesen et al., 2007; Merritt, Alila et al., 2006; King, 2006). Since only 28% of UK households are at present metered for domestic water use such a policy change would have significant impacts on UK households and the water sector in general.

Water metering trials have been undertaken around the world over the preceding 30 years. Gadbury and Hall (1989) report on the establishment of the original UK metering trials of 53,000 households in the Isle of Wight, UK and in 11 other areas of the UK. The "National Water Metering Trials" ran from April 1989 to March 1993 and were set up to assess the practicalities of large scale (Smith and Rogers, 1990; Gadbury, 1989).[4] By 1993 the early results of this trial had shown that the

---

4   Smith and Rogers (1990) report on the difficulties as of 1990 encountered in the Isle of Wight trial, including the higher-than-expected number of cases of multiple properties served by a single service pipe, high incidence of supply-pipe branching before an exterior wall was encountered and electrical earthing issues. Reporting on the Hotwells Trial

average reduction in domestic consumption associated with compulsory metering was found to be 11%, though DEFRA (2006) itself claims savings of between 10 and 15% (and ministers have been known to claim as much as 20%). However perhaps as much as 40% of this reduction was actually the product of better leak detection rather than reduced/disciplined household consumption. The trials also suggested that while compulsory metering had marked effects on *peak* demand with a 30% reduction recorded in peak monthly, weekly, daily and hourly demand (Dovey and Rogers, 1993), they had relatively little effect on *average* demand. Finally, there was evidence, even in this limited trial, that the conservation effects wore off after a short time, suggesting that consumers quickly become inured to the existence of the water meter under the kitchen sink (or in the road outside). Thus, on the UK evidence, the true impact of metering needs to be seen in terms of better leak detection, modest (though likely temporary) reductions in domestic consumption, reduced peak consumption but higher cost and complexity in customer billing and management.

Not surprisingly the metering issue has been widely discussed and examined around the world (ABS, 2006; OECD, 1999; 2003). About two-thirds of OECD member countries already meter more than 90% of single-family houses, although *universal* metering remains a controversial issue everywhere. Selective metering is less controversial, particularly if the public knows that new water resources are scarce or if the metering applies to discretionary water use, like private swimming pools. Metering new homes is also more widely accepted than converting existing ones, perhaps because the issue of compulsion is experienced differently – householders are not "forced" to get a meter, but merely find it increasingly impossible to buy a house without one. Based on work in Argentina, Chambouleyron (2003) points out that without careful regulation of universal metering roll-out only sub-optimal results are likely to be obtained, both economically (for the customers and for the company) and environmentally (with respect to public policy intentions of reducing overall consumption). He proposes that for metering to achieve these objectives a strong regulator is required to manage the transfer of rights to meter from consumer to provider, that the links between metering and tariffs must be clarified and that distinctions ought to be made between *Universal Metering, Optimal Metering,* and *Demand Metering.* The OECD report cited above too suggests that whilst Universal Metering may be non-economic and politically unpalatable, Optimal Metering (i.e. metering of key water users such as large buildings, buildings with pools, etc.) may achieve markedly better results. In a subsequent paper (Chambouleyron, 2004) further argues that decisions about optimal types and numbers of meters can only be made conjointly with decisions about optimal tariffs. Thus far such subtleties have been largely absent from the debate about water metering in the UK.

(Bristol) Bessey (1989) similarly comments that the challenges posed by issues of single pipes serving multiple properties (apportionment) and electrical earthing were significant and unexpected.

Overall we are left with a number of key questions to which policymakers have as yet provided no answers. For example, questions about metering are intrinsically linked to questions about pricing; therefore, metering discussions should not be taken as separate or an add-on. Moreover, since water is unlikely to become scarce in many British regions during the lifetimes of meters installed in the next decade, what is the financial or conservation sense in insisting on universal installation starting in 2010? If one of the purposes of metering is to discipline growing demand for water then we need to consider very carefully the experience in other European countries and in North America where metering is already much more widespread than it is in the UK and the case for a conservation effect is not proven.

Integrated Water Resource Management – The 21st Century

While the TVA is sometimes described, and describes itself, as the first large scale attempt at integrated water resources management, current developments in the IWRM community import values and mechanisms that would have been largely alien to the highly neocorporatist TVA. The first and most important element involves participation in water allocation and development decision-making. In an increasing number of river basins and aquifers around the world, water is already fully or over allocated. A vital task is that of simultaneously reducing (or capping) abstractions and allowing new, high-valued claimants access to water while protecting the needs of the poor and the environment. This problem is, of course, most acute in arid areas of the world. In many arid countries – including Australia, Brazil, Mexico, and the southwestern United States – the phenomenon of formal water markets is being established and maturing. Formal water markets are a central part of a solution to the classic common property resource problem. This solution involves restricting access by potential new claimants and creating incentives for existing claimants to manage the resource instead of overexploiting it, usually by assigning rights to the resource. Government plays several vital roles in this, namely in facilitating and sanctioning the organisation of users into basin and aquifer associations, protecting the legitimate interests of third parties, acting as partners by providing information, and helping monitor and enforce agreements.

If the IWRM principle is adopted, then basin-level systemic management is clearly needed. Every river basin system should be managed *holistically*. This implies an integrated approach to questions of supply and demand management, to issues around competing uses locally, regionally and transnationally. It implies the integration of previously excluded (because "non-expert") stakeholders such as communities, local governments, interest groups such as fishing and boated societies, etc. Accordingly, governments should set up management agencies at the basin and aquifer levels, and international funding agencies should be willing to support and help finance the setting up and strengthening of such agencies.

The creation of such agencies is consistent with two global trends: the demand from stakeholders for a role in managing their resources and the related trend towards greater devolution in all countries. Going back to the historic experience in the Ruhr Basin in Germany at the beginning of the 20th century, a model for participatory basin management has developed. The German model was adapted and scaled up by France in the 1960s. In recent years many countries, including Brazil, Mexico, South Africa, and Zimbabwe, have adopted similar approaches to basin management. A central feature is the integration of participation and the use of economic instruments. It is not by accident that the French basins are governed by parliaments and are known as "river basin financing agencies". Money is the lubricant of accountability in these systems, with most of their revenues raised from user fees and polluter fees and applied to priority public works – many for environmental improvement in the basins.

With IWRM what is called for is nothing less than a new paradigm in water governance, something towards which we have been developing since at least the beginning of the 20th century (this was my argument in Chapters 2 and 3). The EU Water Framework Directive is a specifically European approach to IWRM, with the emphasis on the integration of "stakeholders" onto "River Basin District Liaison Panels" organised spatially around catchment basins. We will have more to say about IWRM and the WFD in the final chapter of this book. For now, interested readers are referred to examples of IWRM operating in Australia and SE Asia by the Murray-Darling Basin Commission and the Mekong River Commission respectively (both of which have good institutional websites).

Concluding Comments

This chapter, and its predecessor, has been, I think, quite challenging. From our earlier forays into the socio-cultural and regulatory dimensions of modern water management (Chapters 1-3) we have now moved through two chapters that are quite "mathematical" and "scientific". I have tried to make the quantitative and scientific aspects of water management as "user-friendly" as possible, but of course there are some technical issues that cannot be avoided or must be treated on their own terms. I cannot imagine, for example, how else to develop a systematic approach to the quantitative aspects of water management except through multiple case studies and exercises. I have however reserved the more complex mathematics of water quality and water allocation management to appendices. Perhaps many of those reading these words did not have much difficulty with the mathematics of water abstraction and water balance statements – to those readers I say "well done"! If, on the other hand, some readers found these passages too complex and laden with abstruse formula, that is not necessarily a cause for concern – the most important thing is to try to get to grips with the ideas behind the analytical tools and methods presented. This is not, after all, a text on civil or hydro-engineering, it is a general introduction to modern water management for people from all walks of life!

Suggested Activities for Further Learning:

1. Appendix 4 contains a more detailed, more technical, exercise in water supply modelling. Working carefully through the exercise will develop skills in abstraction modelling and, as the exercises are fairly involved, numeracy.

2. Consider other large water development institutions such as the US Bureau of Reclamation, the Spanish National Water Plan, and the Department for Environment, Food and Rural Affairs (England and Wales). How do these organisations currently think about water? Do they support large water developments or do they emphasise "demand management"?

3. Attempt a cost-benefit analysis of one or more domestic water conservation technologies – is the scale of their application an issue in assessing feasibility?

4. Evaluate the claim, made by many governments, that universal domestic water metering is an effective way of reducing domestic demand.

5. Try to prepare a water balance statement for your own district, region or nation. Is it possible to begin to see ways of optimising the allocation of water resources between competing users with a water balance model?

Chapter 6

Issues in European Water Management 1:
Water Utility Privatisation in Europe

Introduction

This chapter provides a general introduction in a European context to one of the significant policy challenges to emerge in the last two decades: the privatisation of water supply and sewerage. With the political ascendancy of Reagan in the US and Thatcher in Britain in the late 1970s neoliberal ideas about the virtuous power of unfettered markets to create value and to – as the popular phrase of the time had it – "float all boats" were applied wholesale to nationalised and/or heavily regulated utilities. In time telecommunications, energy, postal services and, eventually, water were either deregulated or sold off lock stock and barrel to private investors. Thus started a roller-coaster process of sell-offs, mergers and acquisitions and horizontal (re)integration that led, ultimately to the spectacular failure of Enron in 2001 and Worldcom in 2002. Whilst both of these cases were American, what many Europeans fail to realise is that Enron was, at that time, moving into the British water sector. Enron CEO Rebecca Marks even proclaimed, chillingly, that Enron would "do" for British water what it had "done" for the American energy sector! Very fortunately Enron sold its sole European water company, Wessex Water, to YTL Holdings, in 2001, less than six months before Enron filed for bankruptcy protection. Other European initiatives, such as Waterdesk.com, an attempt to create, singlehandedly, a market for virtual water futures, were closed down at the same time. However, although the English and European water sectors avoided being seriously affected by the collapse of the likes of Enron and Worldcom, our water sector has now moved into a dangerous phase, with virtually all English water companies now in the hands of merchant banks like Macquarie of Australia and Alinda Capital Partners of the UK. Simultaneously, the big French water multinationals, Suez/Ondeo and Vivendi/Veolia, both of France have restructured their global portfolios over the past few years and are again on the prowl for water sector acquisitions.

In this chapter I present an introduction to water sector privatisation in Europe. Starting with a brief discussion of the "big bang" privatisations of the English and Welsh water sectors in 1989 and key privatisation struggles elsewhere in the world, I move on to the rolling out of the water privatisation programme in central and eastern Europe in the 1990s. Specific case study material comes from the experience of Eastern Europe in the 1990s and 2000s, especially Bulgaria and Poland where the author has conducted primary research since 1992. Along the way it will be necessary to examine the role of multilateral development organisations such as World Bank and the European Bank for Reconstruction and Development and the more recent movement towards "securitising" water, that is treating it as a financial instrument or vehicle, particularly by the investment banks now entering the sector. There is reason to think that, far from slowing in response to the current (mid-2009) "credit crunch", merger and acquisition activity could accelerate as many water companies retain low debt-to-value ratios and significant assets which could profitably be sold off.

Water Privatisation Around the World

Of course the movement to privatise water supply systems has been a fact of life in many developed and developing countries since at least the mid-1980s. As in Eastern Europe, momentum certainly increased in the 1990s, but the roots of the movement lie with certain changes to government policy and corporate structure that took place during the Thatcher/Reagan years. With respect to government policy, the most obvious change involved the application of the neoliberal faith in free markets to optimal resource allocation to the water sector. There was some resistance to the privatisation of water companies, largely because of the obvious social implications of potentially restricted access, but public water suppliers faced a number of intractable problems which made privatisation attractive including:

- Decline of the physical infrastructure at the same time as required quality standards were increasing;
- Inability to raise investment capital without either raising rates or subventions from government (and therefore taxes);
- Conflict between their role as resource developer and regulator (or as Karen Bakker (1996) puts it, "poacher and gamekeeper").

Unfortunately for proponents of state management and development of water resources these institutional difficulties were becoming manifest at the same time as the overall logic of free trade was taking a chokehold over economic regulation the world over. As Barlow and Clarke (2002) point out, this marked the inexorable unfolding of a neoliberal ideology through institutions such as the World Trade Organisation, the North America Free Trade Agreement, and the industry-led World Water Forum (see Chapter 2, page 56). Simultaneously, deregulation and economic globalisation were leading new water multinationals, like Suez and RWE to rise up the rankings of the Global Fortune 500 (98th and 103rd respectively in 2003, rising to 78th and 79th in 2008).

In the context of the Thatcherite/Reaganite 1980s the solution was obvious; indeed these difficulties were taken as proof of the inadequacy of public bodies as allocation mechanisms. Yet nowhere in the western world has gone as far as the UK in privatising water supply and sewerage. The *Water Resources Act* of

1989 set in train the privatisation of England and Wales' ten regionalised water authorities, which was completed in 1991. Since that time the sector has operated as a regulated private market. Cost recovery is generally the primary influence over prices, although there are also price caps in place and the water regulator Ofwat maintains a "market-maker" role in what is essentially a regional monopoly system. Tariff innovation to influence behaviour has however been limited by the low extent of metering of household customers (Day, 2002) and infrastructural investments have, puzzlingly, often not materialised (Bakker, 1996). Moreover, Buller (1996) has found that, in comparison with the still largely municipalised water systems operated in France, the private regional water monopolies in the UK have sown considerable distrust amongst the UK public, making the regulatory functions of central government even more fraught and "political" (cf. recent media frenzies about water rate hikes, "fat cat" boardroom pay, and corruption allegations in UK water companies – e.g. the £1.4 million paid in 2002 to Thames Water's boss Bill Alexander).

Many, many other case studies can be found around the world, many of which (Manila, Buenos Aires, Cordoba, Abidjan, Nigeria, Mexico City, South Africa) are now well-know and well-documented (e.g. Balanya et al., 2005; Barlow and Clarke, 2002; Finger, 2004; Johnstone, 2001; Loftus and Macdonald, 2002). Here we briefly discuss the now (in)famous experiences of Cochabambo Bolivia and Selengor, Malaysia based on recent work published by Olivera (2004), Barlow and Clarke (2002) and Ward (2002). Further discussion of the "big bang" privatisation of the water sector in England and Wales in 1989 will also be provided.

As a consequence of the neoliberal policies implemented since the 1980s, Bolivia has been undergoing intensive economic globalisation and privatisation of basic services and the transport, energy and education sectors. In 1999 it passed a law that, bizarrely, criminalised the patchwork of "traditional" and "cooperative" water supply systems developed by the city's poor by declaring that only a single monopolist could provide water services (Olivera, 2004). When Aguas del Tunari – a joint venture of the US-based Bechtel and the Italian Edison companies – first came to Bolivia, the government promised no more than a 10% increase in water costs as a result of the privatisation. However people were outraged when their water bills showed increases of up to 30%, largely because the concession contract guaranteed a 16% per annum rate of return to the private company *denominated in US dollars*. This meant that as the value of the Boliviano declined in 1999 and 2000 water bills rose to "compensate" the foreign company even though the quality of services had not changed at all. Moreover the company moved quickly after 1 November 1999 to stop all supply services but its own and to issue bills claiming that households were using up to 1,000 litres/day. As a result water bills went through the roof, quantity and quality of services declined and protesters began to fill the streets. During demonstrations in April 2000 more than 100,000 people demonstrated in the centre of Cochabamba incurring military reprisals and then, when resistance did not crumble, government capitulation. The state now runs the water system again.

The Bolivian government was however left with a big problem, and that was that it had signed a 40-year contract with the company, which the popular protests had forced it to abrogate. In February 2002, almost two years after leaving Cochabamba, Bechtel/Aguas del Tunari filed a suit against Bolivia under a bilateral investment treaty, demanding US$25 million in compensation for what it claimed are its lost future profits from the water privatisation scheme and another $25 million in punitive damages (Holland, 2005). Even though Bechtel is a major US multinational, the case is being brought under an investment agreement between Bolivia and the Netherlands since the company thinks that this will aid its case. In January 2006, after four years of secret negotiations and strong-arm tactics, Bechtel withdrew its entire claim and the situation now appears to be at an end.

This case should set off alarm bells about the serious and harmful consequences of investment agreements for the public's right to decide about how it wants its water delivered. The Bolivia-Netherlands agreement is very similar to the Multilateral Agreement on Investment (MAI). And while the MAI is dead for now, similar proposals are part of the proposed Free Trade Area of the Americas (FTAA/ALCA) and numerous regional and bilateral trade agreements. At Cancun in September 2003, the WTO decided to launch its own global negotiations on investment rules. Meanwhile, bilateral investment treaties are being used when companies are unhappy that water privatisation didn't turn out their way. Another recent example: Argentina has been sued over a failed water privatisation plan by Enron's water subsidiary, Azurix. In this case Azurix claimed that it was forced to abandon its 30 year contract to provide water to Buenos Aires after only two years because the government refused to provide additional subsidies when costs of meeting its obligations began to escalate.

In September 2002, the recently corporatised Selangor State Water Department in Malaysia cut the water supply to homes and business premises which had not settled their bills. The company was determined to collect the RM232 million (US$61 million) owed by more than half a million Selangor consumers, which it urgently needed in order to pay back the RM900 million ($237 million) owed to the three water companies that supplied the water. One housewife whose water was cut off complained that the water bill for her house amounted to RM1,700 ($447) over the last three months – about 17% of average income. The last increase in April 2001 saw water tariffs for domestic users increase by 35.7%. Based on the putative "success" in Selengor, the Malaysian government is currently pushing for the restructuring and privatisation of all water suppliers in the country. Primary legislation was passed quickly and without the opportunity for discussion in early 2005, prompting the formation of an anti-privatisation alliance called the Coalition Against Water Privatisation. They face an uphill struggle against a strongly pro-privatisation government backed by similarly bullish international organisations such as the Global Water Partnership.

The privatisation of the water sector in England and Wales in 1989 followed a different pattern. Whilst the arguments for privatisation were the same as in Bolivia and Malaysia, the strategy adopted by the national government was for a wholesale conversion of the 10 public water and sewerage providers and 19 water only companies (established in this form by the 1973 *Water Act*) into private companies in one fell swoop in November 1989. Moreover, this privatisation was, in its first phase at any rate, largely an "internal" affair – unlike the other cases discussed in this chapter where the primary actors are foreign water multinationals, multilateral development banks and weak domestic authorities. Broadly speaking the subsequent 20 years can be divided into three phases: "bedding in" (1989-1995), "consolidation" (1995-2000) and "corporate restructuring" (2000 to the present). In the first phase the new arrangements for water provision and regulation were bedding in, with frequent revisions to the underwriting laws and regulations. New legislation in 1991, the *Water Industry Act* and the revised *Water Resources Act*, completed the new architecture of water services regulation more or less as it currently exists. During this period water prices charged to domestic consumers rose significantly above the rate of inflation (Bakker, 2003; Ward, 1997) and the newly privatised companies began to implement the capital investment programmes upon which privatisation had been predicated. The mid-1990s proved a pivotal period in this bedding in process however as prolonged drought forced water companies in many parts of England to restrict supply, via metering and hosepipe bans. However, as Bakker (2000) points out, by this time English consumers took a very different view of their water providers, expecting them to provide adequate water, in any climatic conditions, as long as people were paying their bills. After all, consumers could not help noticing that come flood or drought, the privatised companies, and their "fat cat" directors, always seemed to make a substantial profit. Thus, in the second phase of the privatisation experience, regulatory authorities clamped down on the companies, primarily by restricting their ability to recoup the cost of required infrastructure investments (which were increasingly about complying with EC environmental and quality Directives).[1] In the third phase, bringing the story up to the present, English and Welsh water companies have increasingly attracted the attention of foreign investors, who have tended to see them as sources of safe, if modest, profits. By 2008 only a small handful of the original two dozen private companies remained in English hands, with some, such as South Staffordshire and Thames Water having changed hands more than once.

These three cases establish the broad parameters of the water privatisation agenda in the 1990s and 2000s. In the first place, there is the coincidence of the realisation of the need for capital investment in a water sector developed "extensively" (and often resource-inefficiently) with the rising ideology that only the private sector can provide both capital and effective corporate governance. This is of course a logical fallacy, created essentially by assuming that the (often pioneer) mass water services managers were necessarily unfit for purpose

---

1 Technically the easiest way to do this within the English and Welsh regulatory systems is to suppress both the "cost of capital" and the overall profit allowed within the periodic "price review" system.

because they were state or quasi-state organisations. Second, from 1979 there was a strong political movement behind privatisation as a way for the state to both wean itself of difficult areas of service and to transfer public capital assets into private hands, often locked in for 25 for 50 years. Where governments were beholden to multilateral development banks like the World Bank and the IMF they generally found that a central condition of any significant aid at all involved the privatisation of utilities, usually to American or European utility multinationals. Third, there was the breakdown, in some places, of the former "statist" view that a certain amount of utility provision (e.g. a certain amount of heat, light, water, education, health, etc.) was a right of citizenship. The ideology of the marketplace meant that increasingly citizens of even social-democratic states were deprived of necessities formerly provided by the state. All of these conditions applied to the postcommunist countries of Central and Eastern Europe, into which the water multinationals rushed even as the dust settled after the Berlin Wall collapsed in November 1989.

Restructuring the Postcommunist Water Sector

The 20 years since the collapse of Eastern Europe's communist regimes have seen dramatic changes in all aspects of these societies, not least in the reorganisation of public utilities like electricity, telecoms and water. This section discusses the general process of postcommunist restructuring in the water sector, involving the break-up of the old national water supply companies, the privatisation of various parts of the new networks of regionalised water companies (e.g. hydro-engineering, drinking water supply) and the introduction of new pricing policies more closely related to the market costs of providing various types of water. Three processes are highlighted which are worthy of scholarly attention: the entry, aided and abetted by strong policy from multilateral development organisations like the World Bank and the WTO, into the emerging water markets of water multinationals (like Vivendi/Veolia and Suez/Ondeo of France),[2] the movement towards redefining water as a commodity which should generate market profits, and the complex part-privatisation of the resource itself. I argue that these processes raise at least as many problems as they solve, particularly with respect to the now very real spectre of unequal access to (especially) potable water and the rise of 'water poverty' even in Britain. It is as yet unclear how water authorities in Eastern European countries will balance the clear need for more market involvement in the water sector against the need to protect and enhance universality of access.

As Table 6.1 shows, the level of water utility privatisation is expected to rise rapidly in all regions outside the "mature" markets of the developed core. Levels of water privatisation are expected to rise more than tenfold in Africa, Asia and

---

2   These companies have been through several restructurings and name changes in the last decade and tracking them can be confusing – see Holland, 2005.

Table 6.1   Levels of water privatisation

Region	Percentage privatised, 1997	Percentage privatised, 2010	Value of privatised market, 2010 (US$, billions)
Western Europe	20	35	10
Central and Eastern Europe	4	20	4+
US	5	15	9
Latin America	4	60	9+
Africa	3	33	3
Asia	1	20	10

Source: Based on Hall, 1999.

Latin America and five fold in the countries of Central and Eastern Europe (CEE). Though levels of privatisation appear likely to remain low by world standards for the foreseeable future, it is worth pointing out that the bulk of this privatisation will involve the largest, and potentially most profitable, urban water supply and sewerage systems. This means that a much larger proportion of CEE populations will be affected by even these modest levels of privatisation. Moreover, the privatisation of the most profitable parts of systems previously integrated at a national level (pre-1989) will de facto mean the *withdrawal* of a potential source of investment surplus which could have been used to redevelop and support the less profitable parts of the network. Put another way, the "cherry-picking" privatisation of large urban water systems, as in Gdansk, Poland, Prague, Czech Republic, Budapest, Hungary and Sofia, Bulgaria not only effectively privatises part of a *common pool resource*, it also necessarily entails a continued drain on the public purse to support the non-privatised (because not commercially attractive) parts of the water system. No current water privatisation deals (and there are a great many variations as we shall see) require concessionaires to take on less profitable parts of the water supply network. Of course the fiscal crises of CEE states has to some extent forced governments to pursue privatisations across the public sector as a way of generating capital for redevelopment of inefficient and outdated communist-era infrastructure (Kindler, 1992).

Another key factor in the accelerating privatisation of water services in postcommunist Eastern Europe are the explicitly pro-privatisation policies of multilateral development agencies, and in particular the World Bank, IMF and the World Trade Organisation. As I discuss below, since at least the early 1990s it has been a key element of World Bank dogma that developing countries must open up as much of previously nationalised economic sectors as possible. Indeed, policy statements such as the 1993 World Bank Technical Paper on *Natural Resource Management* and the 2000 paper on *Natural Resource Management Strategy: Eastern Europe and Central Asia* make it clear that public infrastructure must be privatised to the fullest extent possible (World Bank, 1993, 2000). Furthermore, the strict fiscal policy of its sister organisation the IMF makes it increasingly difficult for even the most basic public services to be funded out of public revenue streams. In both organisations "aid conditionality" is the new watchword. Finally, the World Trade Organisation and discussions around the MAI (Multilateral Agreement on Investment) and GATS (General Agreement on Trade in Services) in the late 1990s make it clear that attempts by governments to protect any area of public services will be aggressively countered. For example, the GATS specifically commits signatories to the WTO to the liberalisation (read: privatisation) of their service sectors, including public utilities. Though they recognise that liberalisation in water services has meant large price increases they nonetheless go on prescribing it to borrower nations. As the journalist John Pilger notes,

> During 2000 water privatisation or full cost recovery was a condition of loans to 12 African countries. The World Bank has promised millions in loans if it privatises its water supply.

And as Holland (2005) and others have noted, by the time the Iron Curtain came down in the late 1980s there was already considerable momentum behind the global push to redefine natural resources, including water, as private market goods.

Taken together, the fiscal collapse of postcommunist states, the parlous state of existing water supply infrastructure, the aggressively pro-privatisation policies of multilateral development agencies and the rise of a small group of water multinationals has made the water privatisation juggernaut almost irresistible. These issues will be discussed in greater depth in the sections that follow, using the best available secondary research literature as well as primary research conducted by the author in the 1990s on water sector restructuring in one Eastern European country, Bulgaria.

Bulgaria

In Bulgaria urban water supply is subject to the controlling influences of at least half a dozen major state organisations. Nominally Bulgaria's water resources are managed by the central government's National Council on Waters (NCW), whose members are appointed directly by the Cabinet of Ministers (Knight et al., 1995; Staddon, 1998). The NCW is charged with developing national water use strategies, allocating water use permits to the various user groups, such as the water supply companies and the national hydroelectric company, setting prices and overall oversight of the country's water resources. The NCW's information and monitoring functions are managed by the National Institute of Meteorology and Hydrology (NIMH), which is itself a research unit of the Bulgarian Academy of Sciences. Research activities related to the water sector are concentrated in the Institute for Water Problems, also a unit of the Bulgarian Academy of Sciences. Legally however it is the NCW which exercises sole ownership functions (on behalf of the commonwealth) with respect to Bulgaria's water resources. The

NCW alone sets withdrawal timetables from Bulgaria's reservoirs including water budgets for industrial, domestic and agricultural sectors.

"Water Supply and Sewerage" companies (Vodosnabdyavane i Kanalisatsiya, or commonly, "ViK") directly manage the water supply infrastructure, including pumping stations, pipes and sewerage networks, throughout the country. ViKs are the most immediate public face of the water sector in Bulgaria. They are the entities that levy charges for water use, though until now the lack of metering at the point of consumption has posed significant difficulties for meaningful price-cost management. Since 1989 pre-existing ViKs have been re-partitioned into 28 regional supply companies and 14 new municipally owned water supply companies, for example in Sofia, Sevlievo and Bourgas. Those ViKs that are organised as municipal enterprises deal directly with central government authorities on matters of water allocation, capital investment planning etc., while the remaining 28 regional public companies are overseen by a special department within the national Ministry for Territorial Development and Construction. The new municipal ViKs even have the power to enter into financial partnerships with foreign firms.[3]

In addition to water supply systems, regional and municipal ViKs are also responsible for the maintenance and management of sewerage and water treatment systems. Though most of Bulgaria's larger towns and cities have developed sewerage systems, less that 5% of the over 1,000 villages are so supplied. Thus, less than 5.25% of all settlements in Bulgaria are serviced by public sewer systems. There were in the mid-1990s only 52 community wastewater treatment plants (WWTPs) with a total throughput capacity of 1.5 million cubic metres/day. Many industrial plants, most of which are still state owned, were also equipped with their own water treatment systems for the immediate treatment of locally produced industrial waste waters. Quantitatively the extent of developed WWTPs pales in the face of actual demand. It is still common, for example, for newer residential neighbourhoods in Bulgarian towns that have not been connected to urban sewerage systems, to dump raw sewage directly into water bodies. Qualitatively, it has been suggested that less than half of treated waters comply with national standards for cleanliness (Staddon, 1996). In addition to the relative paucity of WWTPs, the antiquity of the systems, and the lack of proper management and reinvestment all contribute to inefficient and unsatisfactory waste water management. In the city of Gabrovo in the late 1990s, for example, the city's one WWTP begun in 1974 is still not yet completed, and operates on a "temporary scheme" without much of its planned secondary treatment capacity. In other municipalities WWTP systems have deteriorated rapidly as a function of improper use and lack of capital reinvestment.

---

3 Similarly complex is the South African three tier system where local, regional and national authorities all have active roles in water supply. See the South African "Water Services Bill" *Government Gazette*. 23 May 1997, http://wn.apc.org/afwater/watserbl.htm.

The Sofia urban water supply system is managed by the Sofia ViK, now reorganised as one of the 14 municipally owned ViKs. It manages over 300 km of water mains, a 2,400 km water supply network, over 50 local and central storage tanks and 20 pumping stations as well as wastewater removal and treatment works. Water is drawn from three primary sources:

- The Beli Iskar Reservoir and its catchments providing approx. 1,500 litres/sec;
- The 655 million m³ Iskar Reservoir providing approx. 6,500 litres/sec;
- River catchments coming from the north flank of Vitosha Mountain, located immediately south of the city, providing approx. 130-300 litres/sec.

The 2,700 km long asbestos-concrete water delivery network is in a poor state of repair, being well past its planned 30-40 year planned life span. ViK officials estimate that line losses between source and end user are in the neighbourhood of 50%. In fact, when the possibility of water rationing was introduced in 1993, one water company engineer stated flatly that the system was in such poor repair that the large variations in water pressure implicit in such a proposal would cause widespread system failures.[4]

Most ViKs also receive water from dams and reservoirs operated by the National Electricity Company's Dams and Cascades Division (NEK's Yazoviri i Kaskadi), though the amount and timing of such deliveries varies greatly temporally as well as spatially. One factor influencing NEK water transfers to the ViKs is the current and projected need for hydroelectric power, which NEK is responsible for generating. Water sector installations and equipment is designed and constructed by specialised state firms including, Vodproekt, Energoproekt and Hydrostroi. The various studies and plans for the VK-Rila and other domestic dams, reservoirs and other water developments were produced by one of these state firms.

Until the mid-1990s, 70% of blocks of apartments in Sofiya had only shared water meters, with the subsequent individual usage calculated by dividing the total block usage by the number of registered residents. While costs for water have risen dramatically since 1989 reaching in the mid-1990s the current price of around 10 Leva/cubic metre (approximately 15 US cents/cubic metre, or roughly the average American domestic cost). But this mode of billing by "averages" individual households offers no incentive at all to reduce water use, since charges were essentially fixed independent of actual household use. What arises is a classic common resource problem where it is in no individual consumer's direct interest to conserve water, since the price s/he pays for water is essentially determined without reference to her/his use.[5]

---

4   This is not as unreasonable as it may sound – differentials in pressure, reductions followed by increases, can be even more destructive than the status quo.

5   Again this kind of interpolated water tariff is common throughout the world, being cheaper for the water companies to manage than point of use metering.

Partly because of the fragmentation of institutions concerned with various aspects of water resource development, the kind of carefully controlled multi-purpose development that has characterised western water development since the 1960s has been largely absent in Bulgaria (White, 1968; Mitchell 1997). For example, major reservoir systems such as Belmeken/Sestrimo were designed such that once water is used for power generation it is virtually lost to the drinking water supply system. This fact has led to a great deal of public acrimony and distrust of NEK, with many believing that NEK generates too much hydroelectric power for export to Greece, aggravating the water supply crises faced by the capital. Conversely, it is often the case that relatively expensive treated potable water is being supplied to agricultural or industrial users. Research shows that while overall water withdrawal per capita has been lower in state-socialist nations than in western nations, the latter return a much higher proportion of that water to productive use, resulting in markedly lower actual consumption. Moreover, the condition of water treatment and pumping stations is generally poor with "over one quarter of treatment stations surveyed at the end of the 1980s overtaxed, and one fifth operated with worn out or outdated technology". At the same time, the extensive development of thermal and hydroelectric stations along Soviet-Bloc rivers greatly altered their ability to naturally process organic and inorganic pollutants. Measures of organic pollutants in drinking water, such as observed turbidity and "Biological Oxygen Demand" (BOD_5) demonstrate that overall water quality tends to be unacceptably low (Simeonova ct al., 2003). Kindler (1992) suggests that the economic cost of poor water quality can be as much as 5-10% of GNP. Difficulties in maintaining water quality in Bulgaria are compounded by bureaucratic disagreement and infighting over the establishment of new standards.

At the beginning of the 21st century then the situation facing Bulgarian water authorities and the Central Eastern European water sector is one of a partially completed organisational restructuring in the face of considerable institutional inertia and the reinvigoration of certain vested interests (e.g. Bulgaria's NEK). Moreover Bulgarian consumers are simply unable to absorb significant increases in water tariffs. By the end of the 1990s most CEE countries had not only regionalised the former national water companies, but many, including Lodz and Bielo Bialska Poland, Debrecen and Szeged Hungary, and Vilnius, Lithuania, had also established separate urban water supply companies for the major cities (Hall, 1999). Local government authorities in Sofiya reached a deal to privatise the capital's water supply in late 1999, giving the concession to the International Water Holdings consortium headed by America's Bechtel Corporation. Unlike many other privatisation deals around the world this 25 year concession is in many respects "free-standing"; that is to say that it is designed with *internal* checks and balances rather than depending on a (non-existent) external regulatory structure.[6]

---

6   Interestingly, at the same time Bulgaria also created the possibility for "water associations", a looser sort of privatisation or part-privatisation designed to adapt to

According to the concession contract Sofiiska Voda is to invest a total of US$150 millions in infrastructure improvements over the first ten years of its concession and has the right to increase water rates in proportion to costs of provision.

Elsewhere in Eastern Europe, other potential privatisation deals are at varying stages of discussion and completion. The Romanian government said in mid-1999 that it plans to privatise energy and water. The multinationals are very interested in the privatisation of Bucharest's water system, but no tenders have yet been issued. Tallinn, capital of Estonia, has said it plans to sell a 33% stake in the city's water company, and it is also since the mid-1990s been tendering water supply projects to Western firms. As of mid-1998 Swedish-Finnish-British consortium formed by the public sector companies Stockholm Vatten (Stockholm Water), and the City of Helsinki Water and Waste Water Works, and the UK private company Severn Trent is advising on the restructuring and reorganisation of Vodokanal, the St Petersburg water enterprise, which serves seven million people and employs about 10,000.

The Role of Multilateral Development Agencies

Unlike in Western capitalist countries, characteristic governmental weakness and the endemic economic and fiscal crises have made postcommunist governments peculiarly vulnerable to the privatisation and liberalisation agendas of the major multilateral development the trade organisations. For the countries of Central and Eastern Europe the most important of these organisations have been the World Bank, the International Monetary Fund, the European Bank for Reconstruction and Development, the World Trade Organisation and the European Union. According to various World Bank policy papers, the private sector is to play an increasing role in the development of water services throughout the developing world, and also in the "transition" economies of Central and Eastern Europe. The strategy argues that the private sector is necessary for raising finance, for generating positive institutional change and for ensuring allocative efficiency with scarce resources. Indeed, perhaps the key to the current global fashion for water privatisation is the redefinition, during the 1980s and 1990s of water as an economic rather than a social good. Thus, the International Conference on Water and Environment, held in Dublin, Ireland in January 1992 declared that "Water has an economic value in all its competing uses and should be recognized as an economic good" (International Conference on Water and the Environment, 1992). As a measure of the new importance of privatisation, the World Bank and other international aid agencies are increasingly pushing privatisation though without any common conception or framework. As a result, there is rapidly growing opposition to privatisation proposals from local community groups, unions, human rights organisations,

---

conditions elsewhere in the country where outright privatisation is not attractive. After four were established in Veliko Turnovo and Sliven their legal basis was suspended in 2003.

and even public water providers (Balanya et al., 2005; Barlow and Clarke, 2003; Gleick, Wolff and Chaleki, 2002). In this section I will discuss the policy of the multilateral agencies towards water sector reorganisation with special emphasis on that of the World Bank, as *primus inter pares*.

World Bank policy towards water sector restructuring in developing countries has been repeatedly spelled out in a series of policy documents including the 1993 policy statement "Water Resources Management" (World Bank Technical Paper #12335), the "Water Resources Sector Strategy: Concept Note for discussion with CODE" 7 September 2001 (analysed in detail by Hall et al. 2001), and, with specific reference to Central Eastern Europe, the "2000 Natural Resource Management Strategy: Eastern Europe and Central Asia" (World Bank Technical Paper #485). The fundamental axiom of the World Bank position is that only private capital can properly and efficiently manage public utilities, including water supply:

> Improved water resources management only happens when there are incentives for important actors to make things change. The OED review and the Bank's consultations show that the insertion of the private sector (as operator of an urban water supply or a hydropower plant) provides a powerful incentive to change. The case of the concession contracts in Manila provide a graphic illustration (World Bank, "Water Resources Sector Strategy: Concept Note for discussion with CODE" para 47).

In other policy papers (e.g. #485, 2000, p. 21), the World Bank suggests that this is only one of four "pillars" for sustainable natural resource management (thus "green-mailing" their essential neoliberal "Washington Consensus" agenda):

> A natural resources strategy or substrategy, within this broad framework would have the following elements. Following an assessment of the natural resource base, it would:
>
> • Help client countries assess their natural resource base and evaluate alternatives for sustainable use, taking into account balances between cost-effectiveness, intersectoral, spatial and intertemporal dimensions;
> • Develop plans, investment programs, and environmental assessments for sustainable natural resource management and use, and assure adequate implementation, monitoring, and evaluation;
> • Modify regulations and governance of natural resources in order to assure transparent management and modification in the role of the state and the private sector. Clear rules regarding equitable access to resources, and consensus regarding these, are needed. Decentralized, participatory approaches are often more effective but depend also on transparency in local power structures;

- Modify prices, taxes, and incentives that reflect scarcity and more likely lead to sustainable management. Even where resources are abundant, pricing should reflect the costs of renewal.

While the Bank does not specifically link privatisation to aid conditionality in the concept note, it is hinted at, for example where it says that the strategy will explore ways of making Bank interventions more effective which include:

...becoming more realistic, selective and cost effective...and by being more attentive to issues of the political economy of change *in deciding where and how we engage* (para 36, my italics).

This implies that the World Bank may not be "intervening" – or lending – to countries which decide not to rely on the private sector for water resources. In any case it transpires that in fully 84 of 276 water sector loans given by the Bank between 1990 and 2002 utility privatisation was explicitly required as a condition of the loan (Holland, 2005).

Table 6.2   World bank water sector projects

Project Name	Country	Closing Date	(US$ millions)
Durres Water Supply Rehabilitation Project	Albania	12/31/2001	17.1
Water Companies Restructuring and Modernization Project	Bulgaria	12/31/2002	98
Municipal Wastewater Project	Hungary	12/31/2006	88.9
Bielsko Biala Water and Wastewater Project	Poland	3/31/2005	35.4
Bucharest Water Supply Project	Romania	12/31/2001	50
Water Supply and Sewerage Project	Macedonia	12/31/2006	42.37
ISTRIA WATER SUPPLY and SEWERAGE	Croatia	6/30/2000	143
Istria and Slovene Coast Water Supply and Sewerage Project	Slovenia	12/31/1998	91.1
Municipal Water and Wastewater Project	Albania	N/A	22
Municipal Environmental Infrastructure Project	Croatia	6/30/2006	145.4
Water, Sanitation and Solid Waste Urgent Works Project	Bosnia-Herz	6/30/1999	70
Mostar Water Supply and Sanitation Project	Bosnia-Herz	6/30/2005	13.38
Rural Development Project	Poland	7/31/2004	301.04
Kopacki Rit Wetlands Management Project	Croatia	09/01/03	0.8
Coastal Cities Pollution Control Project	Croatia	N/A	250
Water Supply Urgent Rehabilitation Project	Albania	03/01/03	14.64
Emergency Water Supply and Sanitation Project	Kosovo	1/15/2004	5.92

Source: Based on World Bank Website, 8 July 2002.

Restricting our attention to World Bank policy towards the restructuring of drinking water supply utilities, it is apparent that the primary issues driving the Bank are trade liberalisation and privatisation. With respect to the former the Bank has powerful allies in the European Union, which demands much greater openness to Western European capital as a precondition of EU membership (the "grail" of Central Eastern Europe) and the WTO's General Agreement on Trade in Services (GATS) which demands liberalisation of public services and utilities as a precondition to access to international markets in other areas and avoidance of penalties. The primary vehicles for this global liberalisation agenda in Central Eastern Europe have however been the specific water sector restructuring projects supported by the World Bank, European Bank for Reconstruction and Development and other multilateral and bilateral development agencies. Table 6.2 lists the major such projects supported by the World Bank between 1989 and 2002 together with their intended closing dates and budgeted costs. It is important to note that many of these projects are also tied to parallel projects which may have been funded either by the countries themselves, or through explicitly required public-private partnerships of varying kinds. These projects are all quite similar in structure and purpose and the project description for the Bulgaria Water Companies Restructuring and Modernisation Project is indicative:

> The Water Companies Restructuring and Modernization Project objectives are to assist the government to: (i) increase the corporate autonomy and commercial orientation of the regional water and sewerage companies (RWCs) and make their management accountable to local authorities; (ii) improve health and environmental conditions in urban areas and conserve water resources; (iii) increase operating efficiency and cost recovery in RWCs; and (iv) demonstrate the feasibility and benefits of introducing transparent procurement procedures, efficient contract management, and competition for supply of goods, works, and services. The project comprises two major components: (i) the institutional strengthening window; and (ii) the priority water and sewerage investments window.

Section II of the 1994 Project Appraisal document (World Bank, 1994) further specifies that one of the major objectives of the project is to further commercialise the regionalised water companies with a view to improving access to private investment capital, strengthening the relations between company financial performance and internal decision making, strengthening ability to manage, hire and fire staff and reduce their exposure to legal or political claims through their transformation into limited liability companies (then a relatively new idea in Bulgaria). The project is also designed to promote the privatisation of other parts of the water sector, including Vodproekt and Hidrostroi. Furthermore:

World Bank Strategy Paper for Southeastern Europe, 2000
(World Bank, 2000 The Road to Stability and Prosperity in Southeastern Europe: a regional strategy paper)

Municipal utilities: Private sector projects in this sector are slow in coming and limited in number compared to investment needs and to the large potential for efficiency gains. While there is a gradually growing interest in private sector participation in one form or another, major obstacles remain. In the region, EBRD is working with eight municipalities on the development of projects which involve the private sector in the financing and provision of municipal infrastructure and services, and with a number of municipalities on municipal investment financing without state or commercial bank guarantee. These private projects concern the privatization of the water and sewerage companies in Sofia, Rijeka, Timisoara, Pitesti, Brasov and Burgas, and water sector BOT projects in Zagreb, Bucharest and Constanta. These projects are scheduled for financial closing in or around 2000. They all involve reputable international operators of water and wastewater services. In addition, the EBRD is maintaining an active dialogue on investment financing to municipalities without state or commercial bank guarantees with Dubrovnik, Constanta, Brasov and Pazardzik. Corporatization and commercialization of publicly-owned municipal utilities often lays the groundwork for private sector involvement at a later stage. There are further opportunities in working with municipalities directly, through sovereign structures, where the regulatory and legal framework governing central-local fiscal relations and the provision of municipal infrastructure does not allow non-sovereign or private financing. Especially in countries such as FYR Macedonia, where the EBRD is working with five selected municipalities, this approach remains important and provides the sole entry into this sector.

RWEs, and municipal water companies participating in the Project, will be required to set tariffs at levels that allow for effective operations, maintenance, renewals, and debt service as provided for in the updated tariff setting guidelines to be adopted under the project (Annex 2). Raising water prices over time is expected to result in lower consumption and a general shift from supply – to demand-oriented management.

These are powerful statements of purpose, carrying with them not just the desire to improve water supply for the people of Central and Eastern Europe, but also the radical commoditisation of water services, and perhaps ultimately even the water resource itself. It is telling to observe that in other world regions where this privatisation agenda has been underway for longer, as for example in Latin America, we are now reaching the point where in many countries the water resource itself is now in many places effectively privatised (Holland, 2005). In Chile for example privatising water companies quickly moved to buy up water rights around major cities as a way of hedging against supply deficits (and crowding out competition). Allouche and Finger (2001) have argued that the World Bank programme is so powerful because of its conflation of "market decentralisation" (deregulation) on the one hand and "privatisation" on the other. One effect has been that all roads seem to lead to the same destination, privatisation of the water and sewerage system.

The Entry of Foreign Multinationals and Strategies of Privatisation

There are a wide variety of models of privatisation, ranging from outright privatisation of infrastructure and delivery functions imposed in Britain after 1989 to the contracting out of only specific functions, such as infrastructure maintenance. The fact is that the blanket term "privatisation" covers a very disparate, and even divergent, range of institutional arrangements including:

- **Privatisation:** This term was commonly used towards the end of the 1980s to describe the increase in private involvement; in this paper it refers to the full hand-over of assets (or divestiture) to the private sector.
- **Private sector participation (PSP):** PSP refers to the role that the private sector can play in the delivery of services. There are varying degrees of private sector involvement from service contracts to concessions.
- **Public-private partnership (PPP):** PPP acknowledges the key role that both the public and private sectors have in service provision. The term is becoming increasingly popular as it emphasises the need for partnerships to maximise the benefits which both sectors can contribute. Key implementing mechanisms include "corporatisation" of the water utility and joint ventures, often with functional differentiation.

Full scale programmes of wholesale national privatisation, as happened in the UK and Wales in 1989, have not occurred in CEE, though a number of water supply companies have been effectively privatised through buy-out in Poland, Czech Republic, Slovenia, Romania and Russia, though there are a few examples, such as the 1998 privatisation of the water company supplying Maribor Slovenia (see Table 6.3). In 1992 SAUR privatised Gdansk Poland's water company "Neptun-Gdansk" through the expedient of buying up 51% of that company's shares, and in the late 1990s Vivendi and Anglian Water both effected service provider privatisations through their acquisition of controlling shares in Czech water companies. However, in none of these privatisations has the water multinational also taken private ownership of the resource (and often infrastructure) itself. As Figure 6.1 shows, this implies that privatisation is "asymmetrical", and in fact this sort of privatisation tends to work in the private water company's interest through shielding it from significant risk of infrastructure failure, and through making the public authority an obligatory partner in infrastructure investment and channel for publically-brokered investment. In this regard it is interesting to note that World Bank and EBRD water sector assistance has often taken the form of channelling loan and grant monies through public sector recipients straight into the hands of private contractors (a form of "tied aid" in development). Other common forms of privatisation include the granting of long term "concessions" (usually 20 to 30 years) to operate water services and take revenue (as in International Water's concession in Sofia, Bulgaria) and the establishment of "lease" agreements where the time span is shorter and the obligation to invest on the part of the private operator is reduced (as in the majority of the Czech contracts). "Build-Operate-Transfer" (BOT) type arrangements where the private company agrees to build up substantial water infrastructure in return for access to the revenue stream for a fixed period, often with significant input from multilateral development agencies and the public purse, are less common in Eastern Europe but are appearing in Central Asian countries such as Kazakhstan and Armenia.

The privatised section of the CEE water industry has come to be dominated by a tiny number of companies, based in France, Germany, the UK and the US. The three French companies, Suez-Lyonnaise (now Ondeo), Vivendi (now Veolia), and SAUR, are probably the strongest of the water multinationals with their roots in the operation of water concessions for French local authorities. Hall (1999) comments that "these three French companies…were the only water companies in the world which were private, used to operating across a number of different

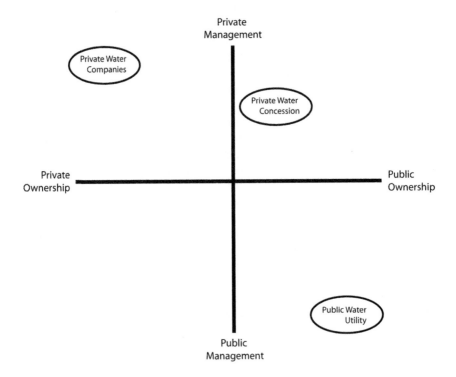

Figure 6.1   Types of privatisation

public authorities, and with the size and capital resources to take advantage of the global fashion for privatisation which started in the 1980s." In the UK Anglian Water and United Utilities have developed a presence in Central Eastern Europe, emboldened by the strong pro-privatisation policy of successive British governments and partly powered by the massive financial reserves of American parents or partners (such as Bechtel Corp as in the case of Sofiiska Voda discussed below). The German multinational RWE, a multinational group operating in a number of sectors, including energy, waste management, telecoms, construction and chemicals, has also developed a presence in Central Eastern Europe's water sector, primarily through its subsidiary Berliner Wasser Beitribe (BWB). And until its spectacular collapse in 2001 the American energy company Enron had been aggressively seeking privatisation opportunities in Europe, primarily through its subsidiary Azurix.[7]

The French multinational Suez-Lyonnaise des Eaux (in 2001 its water arm was rebranded as "Ondeo") is the most established water multinational in Central and Eastern Europe, with concessions in Romania, Poland, Czech Republic, Hungary and Slovenia (see Table 6.3 below). SAUR International (77% owned by Bouygues, a giant French construction company, and 22% owned by Electricite de France, the French state-owned electricity company) uses lease concession and SAUR-controlled subsidiaries to provide water services to Gdansk Poland, the Plzen, Ceske Budejovice, Pribram and Olomouc regions of the Czech Republic and Moscow in Russia. Before the late 1990s, Vivendi had only a small presence in water in central and eastern Europe. However, it has bought control of at least 8 water operators in the Czech Republic in the last few years, including the fast-expanding SCVK, and is now closer to parity with SLE/Ondeo in that country. Both Vivendi and SLE/Ondeo have won concessions in former East German service areas. Vivendi has also entered into joint water provision agreements with two municipal water companies in Romania, Apa Nova Bucharesti serving the capital and Apa Nova Ploiesti, serving Ploiesti.

United Utilities and Anglian Water are the only UK companies that have any presence in central and Eastern Europe. In the Czech Republic Anglian now controls two water companies: Vodovody a Kanalizace Jizny Cechy (VAKJC) providing water to 400,000 residents of South Bohemia and Severomoravske Vodovody a Kanalizace (SmVAK) providing water to 800,000 people in North Moravia. Anglian's expansion into Central Eastern Europe has not been entirely smooth however, with the loss (to apparent mismanagement or underperformance) of an agreement to supply Ceske Budejovice, the largest city in the region (now supplied through SAUR). Surprisingly, the latter company, SmVAK, was won in May 1999 by Anglian after a struggle with Ondeo with 54% of the shares to Suez-Lyonnaise's 45%. The two companies subsequently agreed to cooperate. There have however been allegations of illegal sharedealings and the monopolies

---

7   The story of Azurix is a little-known but fascinating chapter in the story of parent company Enron's spectacular rise and fall.

Table 6.3   Privatised water and sewerage concerns, September 1999

Country	Concession	Company	Multinational	Percent owned	Bought in 1998 or 1999
AL	Elbasan	Elbasan Water	Berlin Wasserbetriebe	–	2001?
BG	Sofia	Sofiyska Voda	International Water	75	1999
CZ	Brno	Brno VaK	Suez-Lyonnaise*	31	
	Ostrava	Ostravke VaK	Suez-Lyonnaise	40	
		Jihomoravske VaK	Suez-Lyonnaise	34	
	Karlsbad	Vodarny Karlovy Vary	Suez-Lyonnaise	49.8	
	North Moravia	Severomoravske VaK	Suez-Lyonnaise	45	
		Severomoravske VaK	Anglian Water	53.4	1999
	Southern Bohemia	VAKJC	Anglian Water	97	2000**
	Ceske Budejovice	Ceske Budejovice VaK	SAUR		
	North Bohemia	SCVK	Vivendi	43	1999
		Aqua Servis	Vivendi (via SCVK)	27.2	1998
		VaK Kladno-melnik	Vivendi (via SCVK)	17.7	1998
		VaK Mlada Boleslav	Vivendi (via SCVK)	22.9	1998
		SVS	Vivendi	100	1998
		Vodospol Klatovy	Vivendi	100	1999
		AQUA Příbram	Vivendi	100	1999
		VOSS Sokolov	Vivendi	100	1998
	Beroun		Anglian Water		
	Pilsen	Vodarna Pilsen	Vivendi	98	
	Prague	PVK	Vivendi/Anglian		2000 ?
ES	Tallinn	AS Tallinna Vesi	International Water	51	2000
GER	Potsdam	Eurowasser	Suez-Lyonnaise		
	Warnow	Eurowasser	Suez-Lyonnaise		
	Kaposvar	Eaux de Kaposvar	Suez-Lyonnaise	35	
HU	Kaposvar	Eaux de Kaposvar	Suez-Lyonnaise	35	
	Pecs	Pecsi Vizmu	Suez-Lyonnaise	48	
	Budapest	Budapest Water	Suez-Lyonnaise/ RWE	25	1998
	Budapest	Budapest Sewerage	Vivendi/Berlin Wasserbetriebe	25	1998
	[Szeged]	[Szegedi Vizmu]	[Vivendi]	[49]	[terminated]
PL	Gdansk	SAUR Neptun Gdansk	SAUR	51	1992
	Bielsko-Biala	Aqua AH	International Water		
	Tarnowskie Gory and Miasteczko Slaskie		Veolia	64	2003
RO	Bucharest	Apa Nova Bucharesti	Suez-Lyonnaise	–	2000?
	Ploiesti	Apa Nova Ploiesti	Suez-Lyonnaise	–	2000?
SL	Maribor	Aquasystems d.o.o	Suez-Lyonnaise/ Steweag	100	1998

Notes: *In 2001 Suez-Lyonnaise water arm was rebranded as "Ondeo". **In 2003 AWG sold VAKJC to Energie AG of Germany.

Source: Table adapted from Hall and Lobina, 1999.

regulator has complained that the multinationals have not taken the need for consumers to benefit seriously enough. More recently Anglian has entered into a contract to provide water to the town of Beroun, west of Prague and was part of the consortium that privatised PVK, the Prague water company.

United Utilities, based in the UK, has been active in Central Eastern Europe primarily through a contract to provide "expertise to improve the quality of water and waste-water services in Bielsko-Biala, Poland and its partnership with International Water Holdings (US) to operate the Sofia water supply and sewerage system" (United Utilities, 2000). This success came after several years of internecine wrangling. In 2000 United Utilities, together with International Water Holdings won the concession to operation the water company, Aqua SA of Bielsko-Biala in southern Poland. In late 2000 UU succeeded in another target market with the acquisition, in partnership with International Water, of a majority shareholding in AS Tallinna Vesi, the company which supplies water and wastewater services for the 400,000 people in Tallinn, Estonia. The joint venture signed a £500 million, 15-year contract to manage the city's water and wastewater operations.

American water multinationals have had much less success in Central Eastern Europe. Primarily represented by Azurix (created when Enron, the recently failed energy multinational, bought Wessex Water, a UK water company, in June 1998) and the US multi-utility Bechtel International through its 50% interest in International Water Holdings. Azurix has found it difficult to establish itself in Central Eastern Europe, but has succeeded in the following ventures internationally:

- It was part of a joint venture, with SAUR and Italgas, which won the Mendoza concession in Argentina in July 1998;
- It bought a bankrupt US contracting firm, Phillips, for its water contracts with some cities either side of the US/Canada border;
- It has bought half-shares of joint ventures in Mexico – Cancun, and IASA – from Aguas de Barcelona and Severn Trent respectively;
- Its first major concession won in its own right was for OSBA, province of Buenos Aires, in July 1999 – a 30-year concession for which it paid $438m.

Its greatest disappointment was failing to win the bidding for Berlin's water supply company BWB, despite offering the highest price. In November 1999 the company fired one third of its workforce because of poor financial results, in March 2000 the company pulled out of a bid to run Bucharest's water system (citing the need to focus on "internal strategic decisions concerning reallocation of capital to priority areas"), and in late 2000 it was absorbed back into Enron, which folded, spectacularly, in late 2001. Azurix's interest in Wessex Water was sold to Malaysia's YTL Holdings in March 2002.

The only other American company to win any foothold in the CEE water market is International Water Holdings (a joint venture of Bechtel Enterprises and United Utilities) which in December 1999 won a 25-year concession to provide water and wastewater services to the city of Sofia, Bulgaria. A new company, Sofiyska Voda, started operation in 2000 and was originally 75% owned by the IWH/UU consortium. An aggressive operator, at the end of 2000, IWH was working on eight projects in Asia, central Europe, Australia and the United Kingdom. By February 2001, IWH had made two more acquisitions: the Tallinn's (Estonia) main water supplier, AS Tallinna Vesi (EUR 42.6m), and the Ecuadorian water supplier Interagua CA Ltda (EUR 10m). In mid-2003 UU bought out its American partner in its Eastern European holdings.

Germany's RWE is one of the largest companies in Europe and the world, a multinational group operating in a number of sectors, including energy, waste management, telecoms, construction and chemicals. In early 1999 it, together with Vivendi, won a 49.9% equity share in Berlin's water company, Berliner Wasser Betriebe, though this was delayed by popular opposition and by a court case was pending which challenged the constitutionality of the sale. RWE has a declared strategy of becoming a "multi-utility", to establish itself as a significant operator in water, gas, electricity, waste and telecoms. It sees the acquisition of a stake in BWB as a major step forward in this:

> With the involvement in BWB the number of customers supplied with water by companies of the RWE Group is rising to 8 million in Germany alone, plus another 2 million in Europe outside Germany...Jointly with BWB, the partners of the consortium intend to take advantage of growth opportunities especially in central and eastern Europe.

BWB's main presence outside Germany is as a partner of Vivendi in the Budapest sewerage company, but it has won a water concession in Lahore, Pakistan been involved in bidding for a number of other water operations internationally in Yerevan (Armenia), Baku (Azerbaijan), Poznan (Poland), Tallin (Estonia), Panama, Chile (in consort with Biwater), and Thailand. For over two years reports have claimed that Poznan is on the verge of privatising its water, and in September 1998 it announced that it would invite tenders, with the help of the EBRD. The next month it was reported that BWB was being considered to form a joint venture with the city council (Orwin, 1999).

The Experience of Privatisation in Central Eastern Europe: Problems and Issues

It is probably accurate to say that the initial "wild, wild west" phase of water utility privatisations in Europe, east or west, is now over. The issues now are not so much about whether or not we allow private involvement in the water sector, but rather what happens now with the configurations of public and private involvements typical in different European states. There are a number of issues that the experience of water sector privatisation to date raise, only some of which have been well-discussed in the literature, including massive price rises and the difficulties caused by underperformance of private water company in the areas of capital investment and social equity. Others are perhaps less well-known, such as the financial complexities that can arise when there are unclear lines of command and control over the water resource, when there are complex deals to channel multilateral development money through governments to the private companies and when water multinationals not unpredictably work to take control of the most profitable urban water supply networks whilst eschewing the less profitable peri-urban or rural networks. Some of these difficulties will be discussed in this section. Finally there is the growing problem of the "securitisation" of water. Whilst Enron/ Azurix (thankfully) failed to get much traction in the European water sector in the late 1990s successive waves of M and A activity by non-traditional capital interests, particularly investment banks, is progressively converting water, or water services into tradeable securities with all the attendant risks so well demonstrated by the current global financial crisis. Three problems in particular need to be highlighted: the inexorable rise and rise of water prices to feed corporate profits, the universal tendency to leverage the relationship with the state to nationalise risk and debt whilst simultaneously privatising profits and, finally, the realisation that private water companies are not necessarily more efficient than public ones.

Most privatisation agreements include requirements that the privatised water suppliers must keep price rises within agreed bounds. In the UK, with a relatively strong regulatory apparatus, these price bounds are negotiated on a five yearly basis through the "Price Review" process.[8] In countries with weaker governments private water companies have often abrogated these agreements, increasing prices excessively. What's more, as in Cochabamba, price rises have been *disproportionately visited upon the poorest sectors of society*. In Trencin, Slovakia and in various communities in South Africa poor customers, unable to pay, have been cut-off, whilst in even in England it required a specific act of parliament in 1999 to prohibit the practice (Havlicek, 2005). In Budapest the private water company demanded high price rises, which the council and consumer groups resisted – but the council is being obliged to make good the companies' losses. Miklos Szalka, vice-president of the municipal maintenance committee, said in 1999 that:

> It is now clear that this kind of privatisation, by 25-year concession of management rights, was a mistake. Budapest City is consulted about price increases, but development finance was expected from the powerful foreign investors. However, the new price increases show that the companies do not want to provide finance for development. The sewerage company has proposed a 25-30 per cent price increase, and the main argument for this is the costs of development…Unfortunately, it is now clear, that these powerful foreign

---

8   Perhaps inevitably this process has proven contentious as, for example, during Price Review 2004 when OFWAT was accused to indirectly boosting corporate profits by agreeing a very generous "cost of capital" allowance.

companies do not want to make investments using their own capital – on the contrary, they take as much money as possible from the country, including their management fees (in Hall and Lobina, 2000).

In June 1999 the council refused to approve a business plan for the water company (Fovarosi Vizmuvek – Budapest Waterworks – 25% owned by Suez-Lyonnaise and RWE), holding up price rises and agreements on directors' pay. Water prices rose approximately 300% between 1994 and 1999 and by 2003 the poorest households were paying about 3% of annual income for water supply and sewerage services (Hungarian Central Statistical Office 2004). In 2006 the company was demanding another annual price rise of 9.8%.

Elsewhere, in the Czech Republic, Anglian's investment in VAKJC has triggered a dramatic increase of over 100% in the price of water and sewerage between 1994 and 1997. Moreover, following the company's acquisition of the majority of the equity shares in 1999, water rates to households increased by 40%, while sewerage rates to households increased by 67%. And when Anglian sold VAKJC in 2003 to Germany's Energie AG, that company turned around the demanded price rises far in excess of inflation. Similarly, a 1998 comparison between the price levels of the restructured municipal operation in Lodz, and the privatised operation in Gdansk, showed that prices were 40% lower in Lodz; 1.53 Zl/m^3 versus 2.41 Zl/m^3.

The simple fact is that during the 1990s and early 2000s companies around the world became more sensitive to shareholder demands for ever higher dividends has had consequences for the performance of contractors in privatised water utilities. Usually water companies have been forced to scale back investment in infrastructure, increase prices to the consumer, or diversify into other more profitable sectors. Sometimes a mixture of all three strategies has been pursued, as in Buenos Aires Argentina where Aguas Argentina has been forced to raise prices *and* cut back on infrastructure investments (Orwin, 1999). Furthermore, private water companies frequently raise investment capital through complex development loan vehicles backed, ultimately, by the public purse. Very fortunately the collapse of Azurix, and then Enron in 2001 saved the sector from suffering the results of the sort of speculative bubble that could have resulted from their stillborn attempts to securitise water.

In the Czech Republic in the late 1990s SCVK's sales of water in its own area have been declining – from 117 million m^3 of water in 1992 to 78 million m^3 in 1997, and an expected 69 million m^3 in 1999. Predictably, the company has compensated for this in two ways. Firstly, prices rises have enabled the company to increase income and indeed revenue in 1998 was 11% higher than in 1997, despite a 5% fall in consumption. Secondly, SCVK has expanded its own activities by buying shares in other Czech water companies. These are now an important part of SCVK's profits. In 1998 the company made profits of Kc115.6m on revenues of Kc2.03billion, but half of the profits were generated outside SCVK's own water distribution area. This sort of in and out running is perhaps inevitable when private utilities adopt the view that satisfying shareholders is more important than providing quality services.

Privatisation of water companies is usually predicated on the assumption that private concerns will perforce be more efficient than public ones. In many cases this carries some grain of truth in it at least insofar as public water utilities have often been financially inefficient, slow to adapt and significantly overstaffed. Regulatory bodies too have sometimes had a difficult time determining the optimum relationship with the sector, sometimes with unfortunate results. Though the UK's water companies complain that the industry regulator, OFWAT, has too much control to require specific investments and to cap price rises, the UK situation is not nearly as bad as that in Caracas, Venezuela where the regulatory bodies were so ambiguously designed and internally fractious that the planned 1992 privatisation of the capital's supply failed utterly. However it appears that there is no clear relationship between ownership type and efficiency (Barlow and Clarke, 2003; Lobina, 2005).

In July 1995, Debreceni Vizmu (Debrecen Waterworks Incorporated Company) was created as a separate entity from the municipality of Debrecen Hungary, with a proper evaluation of assets. The company is now well-established, has financed and carried out major investments, and is operating efficiently. Data shows the financial performance of the company in the last two years of operation (Hall, 1999). Overall, Debreceni Vizmu appears no less efficient than the privatised water companies. The cost of investments has also been lower than would have been the case under privatisation. The relative cost of financing the necessary investments has proved to be much lower under public provision. Investments were in fact determined by Debreceni Vizmu's management on the basis of an assessment of technical needs, rather than on profit maximisation. Twenty-three kilometres of pipework had been finished by April 1997, at a cost of HUF 320m, equal to 40% of the amount Eurawasser would have spent on the same work. The purchase of local supplies rather than French-manufactured equipment, such as meters, has proved more cost-effective. For example, plastic pipes were purchased at 30% less than the French TNCs would have paid, with no transportation costs. As a consequence, the infrastructure had been developed more extensively while the resulting prices were 75% lower than offered by the private concessionaires.

Concluding Comments

New voices are beginning to be heard in the debate over water services provision, and new ideas – good and bad – considered. Among the most powerful and controversial of these new ideas is that water should be considered an "economic good" – subject to the rules and power of markets, multinational corporations, and international trading regimes. In the last decade, this idea has been put into practice in dozens of ways, in hundreds of places, affecting millions of people.

Prices have been set for water previously provided for free. Private companies have been invited to take over the management, operation, and sometimes even the ownership of public water systems. But the trend toward globalisation and privatisation of fresh water is not necessarily a bad thing. In some places and in some circumstances, letting private companies take responsibility for *some* aspects of water provision or management may help millions of poor people receive access to basic water services. However, there is little doubt that the headlong rush toward private markets has failed to address some of the most important issues and concerns about water. In particular, water has vital social, cultural, and ecological roles to play that cannot be protected by purely market forces. In addition, certain management goals and social values require direct and strong government support and protection. Perhaps fortunately the EU has shown a commitment to continued strong public sector involvement in public utilities:

> Solidarity and equal treatment within an open and dynamic market economy are fundamental European Community objectives; objectives which are furthered by services as social rights that make an important contribution to economic and social cohesion. This is why general interest services are at the heart of the European model of society (European Commission Communication 1996, 1).

This implies that EU policy is not inimical to public monopolies in utilities…and this has been upheld in the Amsterdam Treaty of 1999 – all of this implies that whilst European Union policy is generally pro-liberalisation, there are ways of countervailing their global trend even from within erstwhile pro-liberalisation institutions (Conca, 2005; Heretier, 2001). As noted in Chapter 2 another strand of the emerging global level of water governance is the increasingly well-developed and well-networked alliance of anti-privatisation, social rights and environmental groups.

Suggested Activities for Further Learning:

1. Prepare a case study of water services privatisation in a country of your choice.

2. Alternatively, you might prefer to write about the process now known as "remunicipalisation" – that is, the de-privatisation of water services by local or central governments, usually as a result of poor performance or contractual breaches by the private provider.

3. Consider the World Bank and WTO positions on water services privatisation. Water has proven a very contentious subject for privatisation and there are signs that the World Bank at least is softening its stance on privatisation (which it used to insist upon as a matter of course).

4. In recent years water companies have proven attractive to investment banks, which see them as assets with large land banks (which can be sold off to generate a return) and a guarantee of public monies to further prop up corporate profits.

5. In England and Wales, the privatised water sector operates in a simulated marketplace, with Ofwat as the primary regulator responsible for instilling competitive-type discipline into what are still regional monopolies. How well does this system work? You might like to concentrate your attention on a close study of the "price review" mechanism.

Chapter 7
Issues in European Water Management 2: Dams

Introduction

This chapter provides a general introduction to one of the major issues related to water management: the construction and maintenance of large hydroengineering projects including; dams, reservoirs, canals and power plants. By now the reader is accustomed to the idea that water management has, until recently, been dominated by an "engineering paradigm" much enamoured of large supply-side solutions to water management challenges. Case study material comes from a number of locations around the world, and from the European experience over the past century, including such epochal developments as the Elan Valley developments in mid-Wales, the Canal du Midi in southern France, the Franco-era Spanish National Water Plan and the early 20th century hydroelectric developments in central Norway. The chapter finishes with a consideration of the resurgence of dam-building in postcommunist Eastern Europe in the 1990s and 2000s.

Since the primary objective of this chapter is to situate the European experience of hydro-engineering in an appropriate global context, we will begin with a consideration of some of the biggest dam projects ever built: the Three Gorges Project in central China and the Narmada Project in northern India. Though nothing of similar size and scope exists in Europe, this introduction will raise the questions of what exactly constitutes a "dam" structure and the purposes for which they are constructed in the first place. We will then turn to a consideration of the social and environmental consequences of large water projects. All of this establishes a context for thinking through the experience of hydro-engineering in Western and Eastern Europe. Particular attention will be given to the resurgence of hydro-engineering schemes in postcommunist Eastern Europe.

The Geography of Dams in the Contemporary World

On 1 June 2003 Chinese engineers closed the sluice gates on the largest dam complex ever undertaken, the Three Gorges Project, located on the Yangtse River west of Shanghai, allowing the reservoir behind it to begin filling. On 10 June water levels reached a target depth of 135 metres, the minimum for river navigation and for the dam's turbines to begin generating power. The project, first mooted by Dr Sun Yat Sen in 1919, has flooded dozens of towns and small cities, covering scenic spots and historical relics. The government has moved more than 1.4 million people out of the previously densely populated inundation zone. The central dam wall will be 181 metres high, the project will include 26 generators, with a capacity of 18.2 million kilowatts (supposedly capable of meeting 10% of China's mid-1990s electricity needs), create a reservoir stretching 600 km in length and a total cost of more than US $25 billion. According to hydroelectricity generating capacity the Three Gorges Project is 50% larger than the next largest installation, the Itaipu in Brazil, completed in 1983 (Table 7.1). The Three Gorges project, and similarly bombastic ones elsewhere in the developing world, has provoked a worldwide outcry against the environmental and human costs of such development.

There are currently well over 30,000 "large" dams (officially defined by International Commission on Large Dams (ICOLD) as dams with walls over 15 metres high) in the world today, with over half of these in China alone impounding something like 4,300 m^3 of water (Groombridge and Jenkins, 1998). When we turn our attention to the more than 350 "major" dams (over 150 metres in height), then the distribution changes and only a few countries make the grade: the US, former USSR, Canada, Brazil and Japan. Table 7.2 shows the world's largest dams, classified according to the height of their central walls. Topping the list is the Nurek Dam in Tadjikistan, a classic Soviet-era project which impounds a reservoir more than 10 cubic kilometers in volume. Note that of the ten highest dames in the world seven are located in Europe or Eurasia (e.g. Tajikistan) and were built between 1950 and 2000.

These numbers boggle and mind and the reader may well need them to be put into a more user-friendly perspective, so consider the following:

- some of these dams are so big that the mass of their earthworks are more than 8 times the mass of the Great Pyramid at Cheops in Egypt;

Table 7.1   Dams with largest hydropower capacity

Rank	Name	Country	Completed	Capacity (MW)
1	Three Gorges	China	2009 (est.)	18,900
2	Itaipu	Brazil/Paraguay	1983	12,600
3	Guri	Venezuela	1986	10,300
4	Sayano-Shushensk	Russia	1989	6,400
5	Grand Coulee	US	1942	6,180
6	Krasnoyarsk	Russia	1968	6,000
7	Chuchill Falls	Canada	1971	5,428
8	La Grande 2	Canada	1979	5,328
9	Bratsk	Russia	1961	4,500
10	Ust-Ilim	Russia	1977	4,320

Source: McCully, 1996.

- their reservoirs could flood entire countries like Luxembourg to a depth of more than a metre;
- the power from just one of them could provide electricity for a city of a million people.

Fewer than a dozen of these large dams existed before 1950, which implies that the Post-World War II period has been a period of intensive building of dams of unprecedented size. Not only have we crossed an important Rubicon, but we have channelised and dammed it to provide water for drinking, irrigation, industry, power and recreation!

To date virtually all the obvious potential dam sites have now been developed, some intensively. The Columbia River in the US has more than 19 dams, including Canada's largest, the Mica Dam, along its 2,000 km length from northern British Columbia, Canada to the Pacific Ocean at Portland, Oregon (only 70 km of its entire stretch is unaffected by impoundments – see Figure 2.1). All the major river basins in the world have seen some significant level of dam/impoundment development and countries like China and India are moving into a new and yet more rapid phase of dam development.

Of course dams of various sorts have been built for thousands of years. Some of the first may have been built by Chinese engineers more than 3,000 years ago for the periodic diversion of waters from the Yangtse River onto adjacent plains for silt deposition and irrigation. Similar works were developed early on in Sri Lanka, India, Africa (particularly the Nile Basin), Amazonia and elsewhere. But is was really only with the advent of industrial modernity – and the Engineering Paradigm – in the 19th century that nation-builders came to see the permanent diversion channelling or impoundment of fresh water as both an unquestionable good and a national priority. We discussed the genesis of this "engineering paradigm" in Chapter 5.

Table 7.2   World's largest dams by dam wall height

Rank	Name	Country	Completed	Height (m)
1	Nurek	Tadjikistan	1980	300
2	Grande Dixence	Switzerland	1961	285
3	Inguri	Georgia	1980	272
4	Vaiont	Italy	1961	262
5	Tehri	India	u/c	261
6	Chicoasen	Mexico	1980	261
7	Mauvoisin	Switzerland	1957	250
8	Sayano-Shushensk	Russia	1989	245
9	Mica	Canada	1973	242
10	Ertan	China	1999	240

Source: McCully, 1996.

Notwithstanding this long history there are still basically only two types of dam structure: "gravity" and "arch" types. Gravity dams, which are usually the cheapest to build, make up more than 80% of all large dams, and the vast majority of smaller structures. They are generally built across broad valleys near sites where the large amounts of construction material they need can be quarried. They are also technologically the simplest and most obvious kind of design: just pile up enough rock and debris to hold back a natural watercourse. Arch structures, also made from concrete, are limited to narrow canyons with strong rock walls and make up only around 4% of large dams. An arch dam is in form like a normal architectural arch pushed onto its back, with its curved top facing upstream and its feet braced against the sides of its canyon. It is thus a double arch, curved in both plan and side-on perspectives. The inherent strength of the shape enables the relatively thin wall of an arch dam to hold back a reservoir with only a fraction of the concrete needed for a gravity dam of similar height. The Kariba Dam in Zimbabwe is of this double-arched "cupola" type.

Dams are not designed to merely impound water, but rather to fulfil any combination of the following purposes:

- Slow down or change river flow character,
- Divert water to another channel or place,
- Irrigation,
- Drinking water supply,
- Power generation,
- Flood control,
- Navigation,
- Iron out seasonal variations in water availability.

In general parlance, the term "dam" refers primarily to the containment structure that holds back water, but in reality a dam almost always includes a vast array of associated structures and equipment, from turbine generating plants, to spillways, locks, canals, off-takes, draw-down systems and purification and pumping systems. While the idea of a "dam" has an intuitive simplicity, it is in fact a highly complex engineering feat (whatever one thinks about it politically, socially, or environmentally, issues to which we turn later) usually designed to serve multiple purposes. The big structure of the Tennessee Valley (discussed in Chapters 1 and 5) are invariably designed for power generation, navigation, flood control and, increasingly, recreation.

Why Do We Build Dams?

In earlier chapters, I introduced the idea of the "engineer's paradigm", a view of the world in terms of the practical engineering problems its natural form creates. Not surprisingly, this paradigm is well-suited to the perception of dams as solutions to a wide variety of problems. Indeed, through engineering organisations such as the US Army Corps of Engineers, or technocratic development organisations such as the Tennessee Valley Authority in the US or the Murray Darling Commission in Australia, dams have come to be seen as the primary solution to virtually any water or development related problem. Table 7.3 contains a typology of reasons for dam construction, many of which are elaborated below.

Dams as Motors of Development

Dams are not really developed strictly for technical flood control, water supply or power generation reasons. As we have seen, dams are generally conceived as the "jewels in the crown" of national and regional development projects. Thus, the Tennessee Valley Authority conceived of its mission in terms of the integrated hydrological and economic development of the Appalachians region. The dam projects were intended to generate employment through direct jobs in their construction and maintenance and also through the provision of cheap water and power in sufficient volume to support economic redevelopment in what had been traditionally a natural resource extraction and poor hill farm economy. Similarly, the Bhakra-Nangal project in northwestern India, discussed below, was intended as the keystone of the "green revolution" that would fundamentally revolutionise the economies of four Indian states. And finally, the Three Gorges Project is explicitly conceived by the Chinese government as a massive job creation scheme:

> The hydroelectric power generated is to be used in eastern and central China regions and the eastern part of southwest China's Sichuan province. If one figures that every kwh can result in six Yuan in output value, the project may generate 520 billion Yuan in industrial output value every year and create additional millions of job opportunities (Chinese Government, 2003).

National Pride and Dam Construction

There is something thrilling, and deeply troubling, about the way in which political leaders and engineers talk about dams: they are symbolic evocations of national greatness that also celebrate a triumph over a pernicious natural world. From the beginning of the 20th century we encounter countless such evocations. In 1908, none other than Winston Churchill reflected thus on the Nile River:

> On day every last drop of water which drains in the whole valley of the Nile... shall be equally and amicably divided among the river people and the Nile itself...shall perish gloriously and never reach the sea.

Of course the former view – of amicable division – has not prevailed, with periodic belligerent sabre-rattling over the Nile waters by Egyptian, Ethiopian and Sudanese leaders. But the latter view is perhaps the more frightening as it suggests a reduction of a natural phenomenon to a totally engineered, cybernetic "system" – surely a Faustian hubris! Others share this view. Speaking of the Colorado River during the heyday of dam building and river engineering in the late 1960s, Floyd Dominy of the US Bureau of Reclamation declared:

> The unregulated Colorado was a son of a bitch. It wasn't any good. It was either in flood or in trickle.

And Camille Dagenais, former chief of the Canadian dam engineering firm SNC, averred in 1985 that: "In my view nature is awful and what we do is to cure it." The Engineers Paradigm writ large, and we are now beginning to reap the whirlwind sown by this grand hubristic vision!

Dams have also been seen as symbols of national greatness and thereby closely associated with nations' views about themselves. In 1954 India's first Prime Minister Pandit Nehru rhapsodised on the Bakra-Nangal dam:

> What a stupendous, magnificent work – a work which only that nation can take up which has faith and boldness!…it has become the symbol of a nation's will to march forward with strength, determination and courage…as I walked around the [dam] site I thought that these days the biggest Temple and Mosque and Gurdwara is the place where man works for mankind. Which place can be greater than this, this Bhakra-Nangal,[1] where thousands of men have worked, have shed their blood and sweat and laid down their lives as well? Where can be a greater and holier place than this, which we can regard as higher?

Thus, to threaten the viability of the dam is, at some level, to attack the nation itself. It is not surprising that the obverse side of this messianic vision of national becoming has traditionally been ugly and murderous:

> We will request you to move from your houses after the dam comes up. If you move it will be good, otherwise we shall release the waters and drown you all (Indian Finance Minister Moraji Desai speaking to a public meeting at the submergence zone of the Pong Dam[2] in 1961)

---

1   This was the first and largest of the dam projects championed by Nehru. It main wall is 225 metres high, impounds a reservoir 96 km long and contains enough water to supply Delhi's thirst for five years.

2   Located north of Delhi in the sub-Himalaya, the Pong Dam is a standard "gravity" dam constructed of earth and rock and measuring about 100 metres in height and 1.9 km in length. Recent studies have shown that there is now significant leakage from the dam, primarily through seepage underneath and around the structure. Somewhat ironically it has now become a recognised fish and bird refuge and was recently awarded Ramsar Site status in November 2002.

The Role of Multilateral and Bilateral Agencies .

It is undeniable that the number of and size of dam developments around the world would be very much smaller if multilateral and bilateral aid and development agencies adopted a more critical approach to them. The World Bank, Inter-American Development Bank, African Development Bank, Asian Development Bank and other so-called "aid" agencies, as well as private commercial banks, play a major role in the promotion of inappropriate river development projects. By itself, the World Bank itself has wholly or partially funded more than 500 dam projects in 92 countries since its inception in 1948, and its Water Sector Strategy calls for more dams and more private sector involvement, through the now highly contested mechanisms of BOT (build-own-transfer) and PPA (power purchase agreements). Yet, in case after case the benefits have been far smaller than promised, and the costs in terms of money spent, debts incurred, communities uprooted, fisheries and forests destroyed, and opportunities lost have been far greater than imagined. While the Bank claims that its operations have improved in recent years, projects such as the Lesotho Highlands Water Project, the Ertan Dam in China and the proposed Nam Theun 2 dam in Laos reveal ongoing social, environmental and economic problems with its large dam portfolio.

World Bank lending for dams peaked in the late 1970s and early 1980s at a level of more than \$2 billion a year (1993 dollars). Since the mid-1990s the World Bank's lending for large dams has declined significantly. From 1995-1999, the World Bank and its private sector International Finance Corporation made 14 dam-related loans amounting to \$2.3 billion. In the pipeline are five more: two projects involving large dams in China, an energy sector loan promoting hydropower development in Nepal, and political risk guarantees and IFC loans for Bujugali Falls Dam in Uganda and Nam Theun 2 Dam in Laos.

A major reason for the decline in multilateral development bank funding has been the struggle by anti-dam movements across the world. A milestone was reached when the World Bank was forced to pull out of the Narmada Dam project in India in 1993. The Sardar Sarovar (Narmada) project became the catalyst for a global movement of affected villagers, Indian grassroots activists and international groups questioning the World Bank's involvement in large dam projects. Another watershed came in 1995, when the World Bank was forced by citizen action to withdraw from the Arun III Dam in Nepal. Since this time, the Bank has funded fewer dams than ever before, although the relentless rise in energy prices has revised interest in hydroelectric dam projects.

As a result of these facts, international water conservation and development NGOs have, since the 1980s, developed a strong lobby against continued multilateral funding for dam projects. In 1994, 2,154 organisations in 44 countries signed the Manibeli Declaration, calling for a moratorium on Bank lending for large dams and for the Bank to pay reparations to dam-affected peoples. This call was renewed at the first international meeting of dam-affected people in Curitiba, Brazil, in 1997, which also observed that "aid conditionality" generally meant, for poor countries, that they were pressurised not just to build dams, but to do so within the context of a privatised water sector.

Table 7.3   Dam purposes and main construction and operation issues

Purpose of dam	Factors determining dam location	Water level variations in the reservoir	Discharges into downstream river	Water quality	Observations
Hydropower	To maximise the energy produced (depends on river flow and dam height).	Variable on a daily, seasonal or annual basis, depending on production method.	Possibly extremely variable.	Determined by other possible uses of the reservoir.	May result in water diversion from the main river to another catchment.
Public Water Supply	Generally near water consumers/urban agglomerations.	Highly variable.	May be reduced to the legal minimum.	Very good water quality is a primary objective. Must comply with raw drinking water quality standards.	Sedimentation is often an important issue.
Industrial Water Supply	Generally near water consumers/industrial sites.	Highly variable.	May be reduced to the legal minimum.	Good water quality is often very important, however required quality depends on industry type.	Sedimentation is often an important issue.
Fisheries	Variable.	Quasi constant.	Quasi-natural discharge.	Very good water quality is essential.	
Irrigation	Near consumption sites or a channel for water transport.	Seasonally variable. Releases in summer and autumn.	May be reduced to the legal minimum, since water storage is maximised.	A secondary issue.	Sedimentation is a very important issue.
Transport	Dam diverts water into a channel, or acts as a weir, or as a storage facility to allow sluice operation.			Water quality is only an issue if machinery (ships/structures) are affected.	Reservoir may be small or non-existent, but structure can represent a major barrier to fauna migration.

Table 7.3 continued   **Dam purposes and main construction and operation issues**

Purpose of dam	Factors determining dam location	Water level variations in the reservoir	Discharges into downstream river	Water quality	Observations
Recreation	Variable, often a secondary purpose of older reservoirs.	Quasi constant (preferred to maintain beach level).	Possible recreation in downstream river (rafting, canoeing etc.) requiring downstream discharges.	Very good water quality essential – in particular, low trophic state and microbiological counts.	Predominantly a reservoir's secondary purpose.
Flood Control	Variable.	Highly variable.	Dependant on reservoir operating rules.		Requires a significant total potential capacity.
Low Flow Enhancement	Variable.	May vary.	The downstream discharges may reverse the natural flow regime (e.g. flood and low water levels).	Should not adversely affect the receiving water body. Usually designed to enhance dilution. Good quality preferred.	Dam outlets must be operated in different ways.
Spoil Storage	Near to industrial areas.	Often contains little.	Not generally situated on a river.	Impoundment may cause pollution due to leaching.	Type of dam often not relevant to water-related issues.

Source: Leonard and Cruzet, 1998.

Closer to home, the European Bank for Reconstruction and Development (EBRD) is currently funding at least 8 hydropower rehabilitation projects in Russia, Albania, Georgia (2), Azerbaijan (2), Latvia and Slovenia. Other multilateral agencies such as the World Bank (WB), International Monetary Fund (IMF) and EIB are also funding such schemes, described as the "low-hanging fruit" of hydropower development which, if done well, could help make local populations more receptive to further new dam development (UNEP, 2001). Some of these projects will be discussed later in this chapter.

Things (and Dams) Fall Apart: Environmental Consequences

New research commissioned by WWF, the conservation organisation, has warned that dams built with the promise of reducing flooding can often exacerbate the problem with catastrophic consequences, as some recent floods have shown. The research paper "Dams and Floods" shows that dams are often designed with a very poor knowledge of the potential for extreme flood events. Where data does exist it may fail to consider current risks such as increased rainfall due to climate change or increased run-off of water from land due to deforestation or the drainage of wetlands. The loss of these natural sponges for floodwaters within the river basin increases the risk of extreme floods. WWF argues that many of these problems could be avoided if the recommendations of the first ever World Commission on Dams (WCD) were applied to future dam projects.

According to the paper by scientific writer Fred Pearce, lack of adequate information means that dams are often built without adequate spillways to cope with extreme floods. In a 1995 study of 25 Indian dams, World Bank engineers calculated the amount of water that the dams should have been able to release at the height of a flood. In each case, they found the expected floods were greater than those that the dams had been built to discharge over their spillways – two could only cope with one seventh of the expected peak discharge. Furthermore, dam managers often leave it too late to make emergency releases of water at times of very high rainfall and exceptional river flows. This is because their primary purpose is to generate hydroelectricity and provide water for cities, as well as preventing flooding down stream. However as the reservoir overfills they are forced to make releases of water that are far greater and more sudden than flows that would have occurred during the natural river flooding.

Elsewhere in this volume I have already alluded to some other environmental consequences of dam building. One of the most obvious is the disruption of sediment deposition processes in alluvial systems such as the lower Nile Basin. It is well-known that the agricultural productivity of the lower Nile region depends on both the waters of the Nile River and on the nutrient rich sediments which the river brings in its annual floods. Unfortunately establishment of the High Aswan Dam in the early 1960s totally disrupted sediment transport to the lower Nile. One consequence has been increased reliance on artificial fertilisers whilst another has been the rapid shrinkage of dry land in the Nile delta itself and fundamental alteration of the estuary and related fisheries.

Additionally, the fact that reservoirs almost invariably involve the inundation of large amounts of vegetation under water, a situation that leads to the production of large amounts of methyl mercury (a type of mercury far more harmful than the basic substance) through bioaccumulation (Pielou, 1998). Poisoning with methyl mercury can result in neurological symptoms including paresthesias, loss of physical coordination, difficulty in speech, narrowing of the visual field, hearing impairment, blindness, and death. Children exposed in-utero through their mothers' ingestion can be affected with a range of symptoms including motor difficulties, sensory problems and mental retardation. The most (in)famous case of mass methyl mercury poisoning occurred at Minimata Japan in the late 1950s. As a result of growing knowledge about the relations between reservoir construction and methyl mercury poisoning the entire James Bay Cree population in Canada was prohibited from eating their traditional food, fish, after the completion of the James Bay Hydropower Project by the Quebec government in the 1980s.

Ironically there is also evidence that many hydropower dams may release more CO_2 than they save through "clean" power generation. Again this CO_2 release comes as a direct result of the creation, through inundation, of vast amounts of slowly decaying vegetation (Pielou, 1998).

Case Study: Dam-building in the UK

In Chapter 1, I discussed the early history of water engineering in the British Isles, dating all the way back to initial attempts in the Middle Ages to secure safe drinking water supplies for growing urban centres such as Plymouth, Bristol and London. Despite these early works, significant anthropogenic alterations to the hydro-social cycle did not really get underway until the latter part of the 18th century when entrepreneurs realised that, with a little modification, Britain's natural surface watercourses could provide cheap transportation for the country's burgeoning industries. In other parts of Europe canal-building started up to a century earlier (for example the 200 km Canal du Midi in the south of France was completed in 1681), although the "golden age" of canal building is widely understood to have lasted from about 1760 to 1815, after which the age of railways rapidly superseded canals. Later in the 19th century, politicians and entrepreneurs began to realise that whilst the major industrial and population centres were in the central and southern parts of England, the preponderance of water resources were located in Wales and Cumbria. Consequently, in the 1880s and 1890s the British Parliament authorised a number of dam projects for mid-Wales, designed to bring water to the English Midlands and Merseyside. The largest, and most famous, of these is the Elan Valley system, constructed between 1893 and 1904, is comprised of four large reservoirs holding more than 100,000 megalitres. In 1893 the building work began and 100 occupants of the Elan Valley had to move though only landowners received compensation

Table 7.4 Environmental changes associated with dams

Cause of impact	Possible direct effects	Possible indirect effects
Creation of the dam	Creation of a major obstacle in the river.	Barrier to migration for certain aquatic vertebrates, in particular fish.
	Associated construction work (noise, explosions, roads, etc. Population displacement.	Disruption of habitat (e.g. disturbance in the bird nesting season). Increased sediment erosion and temporary effects on river water quality. Population reduction in the vicinity of the reservoir.
	Modification of landscape.	Presence of new water body in landscape (particularly a semi-arid landscape). Cumulative effect on landscape of several dams in the same river basin. Presence of newly-built associated structures (turbine plants, treatment plants) Change in slope gradient – possible increased erosion Creation of a tourist attraction (for recreation). Seasonal population influx.
Reservoir impoundment	Flooding of land.	Habitat destruction – possible loss of rare species Destruction of archaeological and historical features Decomposition of organic material, resulting in temporary eutrophication. Splitting of continuous forested areas in two belts Possible migration barrier for terrestrial fauna.
Presence of a permanent still Water Body	Creation of a still water habitat.	Change from riverine to lacustrine ecosystem Stratification of the water body, with associated changes to the ecosystem water body.
	Creation of a micro-climate.	Increased humidity and attenuated temperature changes upstream of the reservoir.
	Rise in groundwater levels upstream of the Reservoir.	Possible flooding of land (waterlogging) and increased salinisation Changes in groundwater flow regime.
	Effect on bedrock.	Possible induced seismic activity (only in the largest impoundments).

Table 7.4 continued   Environmental changes associated with dams

Cause of impact	Possible direct effects	Possible indirect effects
	Water use.	Change in downstream land use due to the availability of a new water resource (for example, for irrigation). Potential conflicting water demands.
Accumulation in the reservoir	Sediment trapping.	Sedimentation of the reservoir with associated water volume reduction. Reduction of particulate matter in downstream watercourse. Leaching of nutrients and other substances.
	Nutrient enrichment, causing eutrophication.	Evolution of ecosystem. Appearance of water detrimental to recreation uses – toxic algae Increased water treatment required for drinking water supply.
	Atmospheric acidic deposition.	Acidification of reservoir – low pH and effects on ecosystem.
	Chemical pollution.	Accumulation of pesticides, heavy metals and other micro-pollutants.
	Biological pollution from human or animals.	Possible presence of pathogens.
Reservoir Operation	Periodic emptying.	Impact on downstream ecosystem (sustained high discharge flows, water quality due to variations in a stratified reservoirs) Choice of emptying period may be limited with a narrow bottom outlet Possible clogging of downstream banks if no sediment management rules enforced.
	Water level variations in reservoir.	Modification of shoreline ecosystem Effect on landscape of bare rock shoreline.
Controls on upstream catchment	Legislation, regulation or education to reduce sedimentation or nutrient loads to upstream river.	Changes in catchment land use. Alteration of fertiliser application practice Installation of wastewater treatment plants Improvement of upstream river water quality.

payments. Many buildings were demolished, three manor houses, 18 farms, a school and a church (which was replaced by the corporation as the Nantgwyllt Church). In July 1904 King Edward VII and Queen Alexandra opened the Elan dams and water started flowing along 118 km of pipeline to Birmingham. The most recent 25 year Water Resources Management Plan calls for twinning the pipeline from Elan Valley to ensure a reliable supply to the English Midlands through 2030.

Although many reservoirs were constructed to supply cities in South Wales (Cardiff and Swansea especially), the issue of diverting Welsh water to slake English thirst proved quite powerful in stoking nationalist sentiment in Wales (Fishlock, 1972). Ironically, it was a relatively small project, at Capel Celyn in Denbighshire, that sparked the most violent reaction, including several bombings and the enduring nationalist slogan "Cofia Dryweryn" – "Remember Tryweryn" (the name of the valley in which the reservoir was built). Capel Celyn was a small village in the quiet Tryweryn valley in the hill country between Bala and Ffestiniog in North Wales until 1956 when a majority of (English) MPs voted to flood the community by authorising the construction of the Llyn Celyn reservoir to provide water for Liverpool and The Wirral. Putting the bill to the Lords in 1957 Minister for Welsh Affairs Henry Brooke stated, emotively:

> Water shortages might occur in the next few years on Merseyside and in South-West Lancashire. I cannot believe that preservation of the Welsh way of life requires us to go as far as that. I cannot believe that the Welsh people, of all people, want to stand outside the brotherhood of man to that extent.

The dam was opened in 1963 and was subsequently the subject of sabotage campaigns including bombing with stolen gelignite. The bomb attacks on water pipelines and government offices went on for more than four years, but the reverberations of Tryweryn are being felt still. "'Tryweryn', wrote Gwynfor Evans, the Plaid Cymru (Welsh Nationalist Party) leader between 1945 and 1981, 'will become a word of fateful significance for Wales. It may become as well known a verb as "Quisling" has become as a noun.'"

Unlike England and Wales, Scottish dam building has largely been a story of hydroelectric power generation for industrial development. The Kinlochleven and Lochaber hydroelectric schemes were completed in the 1920s in order to provide cheap power for British Alcan's aluminium smelters in the Highlands (and are thus also about local economic development). Reservoirs used primarily for water supply include Glencorse, Threipmuir, Harlaw and Rosebery, all built in the mid-19th century to secure water supplies for Edinburgh.

The compact geography and hydrology of North Ireland has meant that little reservoir construction has been necessary, although by 1930 the Spelga and Silent Valley reservoirs had been built to supplement supplies from Lough Neagh.

In the 1990s and 2000s declining relative water availability (more unpredictable rainfall plus increased population) has meant that hydro-engineering has been put back on the agenda. Adding pressure for further development is the Environment

Agency's "Restoring of Sustainable Abstraction" process (begun in 2005), which is systematically reviewing all water abstraction licences with a view to reallocating water resources for environmental protection where necessary. This has triggered some renewed interest in the idea of a "national water grid" (perhaps like the one in Spain – see Swyngedouw, 1999) bringing water from the relatively wet north and west across the Pennine Mountains to the drier south and east of the island. However much intuitive sense such an idea might have, both the privatised water companies and the UK government are currently opposed to any such plan on environmental, economic and practicality grounds. Instead, as we shall see in the final chapter in this book, all major stakeholders are placing more emphasis on interconnections between adjacent catchments, water efficiency and, especially, better demand management. There are however a number of plans for reservoir extension or new dam building, most (in)famously the Abingdon reservoir being proposed by Thames Water to augment supply in the Thames Basin. This reservoir, currently part of Thames Water's 25 year Water Resources Management Plan, could incorporate a bund wall structure up to 25 metres high and impound up to 25 billion litres of water (making it the second largest reservoir in England after Kielder).

Small and Medium-sized Dams in Eastern Europe

Government dam projects were one of the primary catalysts for the upsurge in popular demonstrations in the communist countries of Central Eastern Europe in the middle and late 1980s. In October 1988 Hungarians took to the streets to protest against the continued construction of the Gabcikovo-Nagymaros dam project (introduced in Chapters 2 and 3). In the same month Bulgarians came out onto the streets to protest against the highly contentious "Water Supply Rila" project that would extend the capital city's water supply feeders deep into the Rila mountains 100 km south of the city (Staddon, 1998). Yet, in the first instance, the popular demand was not for an end to Communist rule, but an end to dam projects that despoil the environment. As studies of environmentalism in Eastern Europe have shown, communist authorities often tended to tolerate such public outcries provided they were aimed at something ostensibly "non-political", like the environment. Yet one result of this anti-dam campaign was that it helped build confidence among Eastern Europe to speak out against, and ultimately overthrow, their Communist rulers.

More than a decade after the electrifying events of October and November 1989, dams are back on the agenda – not perhaps the large dam projects that have occupied our attention so far, but a myriad of smaller projects that may be, cumulatively, *more* damaging than just a few large projects. Punyas and Pelikan (2007) estimate that as much remains to be developed as has already been developed in the formerly communist countries of Central Eastern Europe – perhaps as much as 15 TWh/yr. Sternberg (2008) further notes that dams are akin to "solutions in search of a problem".

The World Wildlife Fund (WWF) has warned that without concerted action at the EU level, the decline in Europe's rivers would continue. Europe has only one large free-flowing river system left untouched by dams: the Tornedlven on the border of Sweden and Finland. 80% of rivers in Austria and 70% in Sweden have been damaged by dams for hydro-power. Loss of river and floodplain habitats has created a dire situation in which natural floodplain forests along the Danube are now only 4% of their original size. Nearly all lowland floodplains in Spain have been lost, and the country has embarked on a new round of dam building (Swyngedouw, 1999). The dense network of dams on Europe's rivers is also hindering natural capacities to flush out water pollutants such as nitrates and phosphates. One in two rivers in the UK and are still "heavily" polluted with phosphorus while one in ten rivers in Belgian Flanders, Denmark and the UK are "very heavily" polluted with nitrogen. Major infrastructure projects are still being developed: a large scale water transfer scheme from the Ebro in Spain, damming and river engineering projects for the Vistula in Poland, and extensive new waterways in Central and Eastern Europe (WWF, 21 September 1999).

By the end of the 1990s it was already apparent that postcommunist Eastern Europe's new democratically elected governments were jettisoning the inconvenient environmentalist baggage that had been so crucial during the turbulent period of regime change in the late 1980s and early 1990s. Green parties found themselves increasingly shunned by the coalitions that ruled in countries like Poland, Bulgaria and the Czech Republic and their electoral support also shrank as hard-pressed populations focussed their efforts on economic survival. Where greens remained within government, it was often under the guise of something else entirely (e.g. their credentials as democrats or scientists) or as very moderate greens, whose politics was not at all inimical to the neoliberal economic agendas of "shock therapy" and mass privatisation. In this context it is perhaps not surprising that CEE governments have increasingly returned to the sorts of hard environmental engineering projects so favoured by their communist predecessors. Power plants (including nuclear), mining developments, water reservoirs and hydropower plants have all returned to the political agenda. Moreover, as in other developing nations, these projects are actively encouraged by multilateral development agencies such as the World Bank and the IMF as the perfect vehicles for the sort of "tied aid" that often benefits the donor countries at least as much as the recipients. Yet, while much attention has been paid to big dam projects such as Gabcikovo-Nagymaros, relatively little is known about the proliferation of small and medium-sized dam projects whose cumulative impact may in fact be far greater than the few large dam projects underway.

EBRD is funding at least 8 hydropower rehabilitation projects in Russia, Albania, Georgia (2), Azerbaijan (2), Latvia and Slovenia. Other multilateral agencies such as WB, IMF and EIB are also funding such schemes, described as the "low-hanging fruit" of hydropower development which, if done well, could help make local populations more receptive to new dam development (UNEP, 2001).

Bosnia and Hercegovina

In July 2000, Austria's Hydropower-Tyrol and Bosnia's Inotrade-Sarajevo companies reached an agreement with the government of Bosnia for a 20-year design, build, operate, and transfer (DBOT) project with the Srednjobosanski canton authorities (in central Bosnia) and a purchase power agreement with electric power utility Elektroprivreda Bosne i Hercegovine. The consortium plans to invest US$6.03 million to construct four hydroelectric projects in central Bosnia. This marks the first investment by a foreign company in Bosnia's power sector. Three of the plants (Prokoska, Jezernica 1, and Mujakovici) are to be located on the Jezernica River. The fourth, the Botun project, is to be installed on the Kozica River. The four "run-of-river" plants, with a combined capacity of 3.8 megawatts, are expected to begin operating within two years.

Bosnia is also negotiating with the World Bank for a $30 million loan for a project to rehabilitate several hydroelectric power plants, the so-called "Power 3 Project", among the 25 dam structures already in existence in the country. Ten other dam projects are at initial design/development stages.

Bulgaria

The complex relief structure and the small territory of Bulgaria do not create conditions for big rivers. Most of Bulgaria's rivers in Bulgaria spring from its high mountains and flow into the Black Sea (predominantly through the Danube River) and the Aegean Sea. The catchment basins of these rivers are generally small, the biggest one is the Maritsa River basin at 21,000 km^2 and the longest one is the Iskur at 368 km. Natural lakes in Bulgaria are comparatively few, the biggest being located along the Black Sea such as the lagoon lakes of Alepou, Arkoutino, Pomorie, and the firth lakes of Beloslav, Bourgas, Varna, Shabla. Most numerous (more than 360) are the high-mountain alpine glacial lakes in the Rila and Pirin. mountains. All the lakes and the swamps along the Danube have been drained with the exception of Sreburna Lake which is a part of a reserve. Numerous dams have been built as parts of hydropower or water supply systems – Iskur, Arda, the Batak Hydropower System, Dospat-Vucha, Belmeken-Sestrimo, and also about 2,000 small dams.

Bulgaria plans to build six new dams starting in 2001 to overcome water shortages caused by the continuing drought, at an estimated cost of US$22 millions (REC, 2001) There are also plans to upgrade and expand numerous hydroelectric facilities. Plans to rehabilitate the Gorna Arda facility in southeastern Bulgaria ran into problems in 2000 when Turkey's Ceylan Holding was forced to withdraw from the project because of financial difficulties. In November 2000, Bulgaria issued new tenders for upgrading the Gorna Arda complex and has announced that state electricity utility Natsionalna Elektricheska Kompania (NEK) will now assume Ceylan's 50% stake in the project. Construction on the three-dam, 170-megawatt cascade project in the Rhodope mountains in Bulgaria's southeast have been repeatedly delayed after Ceylan agreed to a joint venture with NEK in November 1998. As of mid-2002 the problems with the Ceylan holding had yet to be resolved.

Potential Italian investors want to bundle Gorna Arda together with the pre-existing Dolna Arda system of three dams generating 270 megawatts. Once construction begins, it is estimated that Gorna Arda would be completed within six years. Specific funding has also been made available by the World Bank to rehabilitate the Belmeken and Chaira hydropower systems (Project P008316, US$93 millions), including the construction of the controversial Iadenica Dam.

Plans to privatise several of Bulgaria's hydroelectric projects have not run smoothly. In 2000, the country's Privatization Agency (PA) had to ask bidders to resubmit their bids after senior politicians accused bidding companies of corrupting the sale process. The accusations were levelled against companies that had submitted bids for the Pirinska Bistrica and Sandanska Bistrica cascades, which are considered two of the most attractive of the 22 hydroelectric power assets, all with less than 25 megawatts of capacity, being sold by the PA. By July 2000, the PA had sold six of the offerings, including the three-dam Sandanska Bistritsa cascade to the Czech Republic's Energo-Pro and the two-dam Pirinska Bistrica cascade to Bulgaria's Pirin 2001.

Other water development plans include new water supply dams for the Dobrouja, Central Balkan and Rhodope regions, for which the Bulgarian government is attempting to secure Kuwaiti investment capital (Sofia Echo, 17 May 2002).

Czech Republic

As elsewhere in postcommunist Eastern Europe, Czech government and water industry authorities have proposed new reservoirs of all size for all purposes. Dams have recently been proposed on the upper Elbe at Hrensko, Boletice, and Strekov and, in the aftermath of the 1997 floods, elsewhere in Bohemia and Moravia. In late 2006 the government of Vaclav Klaus (who is well-known for his anti-environmentalist opinions – See Pavlinek and Pickles, 2004), published a report listing more than 200 potential dam sites around a country roughly half the size of England (where one proposal is instantly contentious!). One development is particularly well-advanced and will inundate the village of Nové Heřminovy, located in the northern part of the country.

Yet, anti-dam organisations appear to be stronger in the Czech Republic than elsewhere in Eastern Europe and have made inroads into popular understanding of alternatives to dam-building. For example, the Ecological Institute "Veronica" plays the role of a secretariat for the Union for the Morava River civic association, an organisation established in the 1990s to counter new dam proposals in the Morava catchment. A recent study entitled "An alternative ecological proposal of flood-preventing measures in the Morava and Bečva catchment areas" was designed to demonstrate that it is possible to ensure the required protection of settlements in river basin floodplains of the Morava catchment area against flooding without having to build new retention reservoirs. In order to catch and retard the high-flood-water wave in the countryside and river basin floodplains, it has been suggested to use natural elements in such a way that the changes in water run-off situation made could lead to restoring the landscape to an almost natural condition. Similarly, Czech and pan-European initiatives such as "Salmon 2000" have had an effect on river management along the Rhine and Elbe Rivers.

Former GDR

After reunification, the German government proposed a series of new dams on the Elbe River system. European river activists in Europe defeated a scheme of navigation dams proposed for the Elbe River, and convinced the German government to instead improve existing navigation channels, thus leaving the upper portion of the river unharmed (Epple, 1996). The federal Ministry of Transport planned a series of river-regulating dams and channels to facilitate navigation for wide-gauge barges, a plan which threatened the Elbe's ecological balance. In response, environmentalists from Germany and all over Europe came forward to defend the river. Opponents soon managed to reduce the proposed scheme to only one dam on the Saale River, a tributary of the Elbe, and to the strengthening of the "buhnen" (jetties) used to channelise the river for large barges.

The 1,091-kilometer-long Elbe River, which flows through the former no man's land that once separated East and West Germany, came out of the shadows with the reunification of the country in 1990. The river had been badly polluted by East Germany's heavy chemical industries (the huge "Kombinats" which disgraced the land), but the riparian ecosystem was surprisingly intact, because urbanisation along the river was minimised by its status as a dangerous border zone. Consequently, large areas remained wild, supporting a rich flora and fauna and the largest stretch of flood plain forests remaining in Europe.

Finally, on 5 September 1996, the four largest German NGOs and the Ministry of Transport came to an agreement: the Elbe river will be closed to all commercial traffic between Lauenburg (upstream from Hamburg) and Magdeburg as soon as existing canals are widened for wide-gauge ships. Part of this approximately 300-kilometer-long portion may be added to the existing 78-kilometer-long UNESCO Biosphere Reserve – the Mittlere Elbe Biosphere Reserve. Located around Dessau and Torgau, where American and Russian troops invaded Germany during World War II, the expanded Biosphere Reserve would become the longest protected section of river in the world.

The next step for environmentalists will be to protect this portion of the river, the upper Elbe. But the long-term objective is to change the European policy of enlarging rivers for wide-gauge barges. The Rhine and the Altmühl, once living rivers, have been turned into dead waterways from this policy, and French authorities now mean to do the same with the Doubs River to complete the Rhine-Rhône canal. However, this first victory on the Elbe, coupled with the strong opposition to the Rhine-Rhône canal, indicates strong support for protecting rivers from ill-conceived development projects.

Latvia

The European Investment Bank (EIB) is lending ECU 6 million for rehabilitation works and dam safety improvements for the Daugava hydropower plants. Investments comprise rehabilitation work on the generators as well as improvements to the dams on the Daugava River in central Latvia (EIB, 1996). The funds are going to Latvia's state-owned electricity utility Latvenergo to improve the overall efficiency and maintain the capacity available from the hydro power plants. The financing will therefore assist the company to continue to act as a peak-demand supplier in the country and in the region. The investments will also help extend the operating life of the plants, thus delaying the need for major new generating capacity elsewhere. The project is also being supported by the European Bank for Reconstruction and Development (EBRD).

Macedonia

Macedonia's electric utility, Elektrostopanstvo Na Makedonija (ESM), is planning a project to rehabilitate six of its largest hydroelectric plants: Vrutok, Vrben, Raven, Spilje, Tikeves, and Globocica. Most of the plants are more than 40 years old, and their continued operation is considered vital for ESM. The six plants generated 92% of Macedonia's hydroelectricity supply in 1999. The total value of the project is estimated at $52 million, and it is slated for completion in 2004 (although it is already running six months behind schedule). In addition to reconstructing turbines, generators, and transmission facilities, replacing turbine circuits, and repairing transformers, total generating capacity is to be increased by 31 megawatts. Loans from the World Bank and the Swiss government, along with a grant worth $0.66 million from Japan and $12.1 million from ESM's own funds, will be used to complete the project.

There are also plans in Macedonia to develop the Chebren hydroelectric dam on the Crna River. Originally, the project was supposed to consist of two cascade dams, Skocivir and Galiste, along with the Chebren dam, which was to work with the Tikves hydroelectric plant to form the Crna system. However, experts determined that the scheme would be unprofitable in that form. The government is now negotiating with an Austrian consortium led by Alpine and including Verbundplan and VA Tech Elin for a scaled-down version of the project in which only the Chebren reservoir would be constructed, thereby allowing the project to return a profit. If an agreement is reached, the project could be completed within five years.

Plans also exist to divert the Anska River into Dojran Lake, which, like the Aral Sea, is slowly disappearing owing to overexploitation (Rivernet International News, 16 October 2001). The lake currently has an area of 42.5 km^2 and is 1.5 metres deep, 3.5 metres below its optimum level as stipulated in a 1956 bilateral agreement with Greece.

Poland

Poland has 174 class I to III dams, and more than 600 class IV (retaining wall less than 10m) dams (Dams Monitoring Centre, 2001). Unfortunately many of these dams are in poor condition, either due to poor original construction or to advanced age (more than 50 years old). The 69 large dams in Poland (ICOLD database) reveal that the country has more multi-purpose dams than the world average.

According to the WWF, Poland risks delaying its membership of the European Union if it goes ahead with two controversial river engineering projects (WWF, 2000). The two projects are:

- A proposed new 270 million Euro dam cascade on the Lower Vistula at Nieszawa, that would irreversibly damage one of Europe's last big almost free-flowing rivers and some of Europe's most outstanding landscapes and nature;
- The 2,500 million Euro "Odra 2006" project involving two new dams and major river engineering works for shipping, flood defense and water quality improvement.

The government is proposing to build a series of seven dams (called a "dam cascade") along the 394-km length of the lower Wisla, each reservoir extending back to the previous dam. There is already one dam on the river (World Rivers Review, 1998). The proposal also includes an east-west waterway linking with the River Odra in the West, and several more new dams on the Odra itself. Project promoters claim the dams will solve Poland's water supply problems, reduce floods, supply electricity and improve navigation. But most of these claims are unrealistic or even contradictory:

> The Nieszawa dam project is in breach of Environmental Impact Assessment, the Access to Environmental Legislation and the Birds and Habitats Directives as well as being contrary to the Amsterdam Treaty commitment to sustainable development" commented Tony Long. "Poland can put its environmental record back on track for early EU membership by cancelling these projects and working with the Commission and NGOs like WWF to implement greener and more cost-effective alternatives.

The completed project would not only destroy the unique dynamics of the river and habitats of European importance for breeding and migrating birds but also prevent any environmentally-based development. Furthermore, it could also pave the way towards construction of the big scale east-west waterway that is proposed for European Union (EU) funding as part of the TransEuropean Transport Network. This would devastate the Oder, Warta, Vistula, and Bug rivers, and their associated wetlands (WWF, 6 July 1999).

The European Commission has criticised Poland many times for not making enough progress in adopting EU environmental law. During a visit to Poland in March Commissioner Wallström expressed her concern about the impact of the proposed Nieszawa dam. The Commission stated (in response to an oral question from former Danish Health Minister Torben Lund MEP) that it ruled out EU funding for the Nieszawa dam but has confirmed funds could be available for alternative schemes. Most EU countries are moving away from big dam projects towards more cost-effective and eco-efficient solutions.

The dams are opposed in Poland by many organisations including WWF, the World Conservation Union, Klub Gaja Eco-Cultural Society, Polish Society For The Protection Of Birds and Angling World Magazine.

An example of dam-induced flooding comes from Poland where the 1997 floods covered an area only half as large as that in 1934 yet three times as many buildings were flooded, 38 times as many bridges and 134 times more kilometres of road (World Commission on Dams, 2000).

Romania

In Romania there are a number of hydroelectric projects that have not been completed but could substantially increase the country's installed electric capacity. The country has disclosed that there are at least 12 partially built hydroelectric projects in Romania that require foreign investment to finance their completion. In 2000, the US engineering company Harza and Romania's hydropower producer Hidroelectrica signed an agreement to complete and jointly operate one of the plants, the 54-megawatt (reduced from 155 megawatts after an optimisation study conducted by Harza) Surduc-Nehoiasu hydroelectric plant on the Buzau River. Construction on the $60 million project should be completed by 2004.

According to the ICOLD, in the mid-1990s there were 39 dams more that 15 metres high under construction in Romania (Jan Veltrop, "Benefits of Dams."*USCOLD Newsletter*, March 1995, p. 3).

Russian Federation and CIS

In Russia, the FSU's largest economy and electricity consumer, hydroelectricity accounts for about 20% of total electricity generation or approximately 43,000 megawatts. Almost three-fourths of Russia's hydroelectric capacity is represented by only 11 power stations with more than 1,000 megawatts of capacity, including the 6,400-megawatt Sayano Shushenskoye facility in Khakassia, the 6,000-megawatt Krasnoyarsk facility in Krasnoyarsk province, and the 4,500-megawatt Batsk project in Irkutsk province.

Hydroelectricity makes up more than 80% of the former Soviet Republic of Georgia's electricity generation. The country has not yet fully exploited its hydroelectric resources and has an estimated 100 billion kilowatts of potential hydroelectric capacity. An Austrian-Georgian coalition of Strabag, Verbundplan

GmbH, ABB Kraftwerke AG, and Lameyer International GmbH plans to invest up to $500 million over the next several years for hydroelectric projects. The 250-megawatt Namakhvani and the 100-megawatt Zhoneti projects are planned for construction on the Rioni River, and the 40-megawatt Minadze station is to be constructed on the Kura River. The EBRD also has approved a $39 million loan to refurbish the Inguri Hydroelectric dam, the largest in the country.

Water supply dams are also going ahead in the Russian Federation. In late 1998 the government of the Republic of Bashkiria approved the construction of the Yumagusinsky (Ishtungansky) Dam in the middle Belaya River (*ERN Newsletter*, 1998). This is an integrated water supply and hydroelectric project which is planned to provide up to 127 million kWh of electricity annually when fully operational. The origin of the project goes back to the mid-1970s when a project named "Ishtungansky Dam" was prepared and considerable resources where spent at the erection of the workers village "Yumagusino", when several quarries where laid out and the construction work had begun. At that time it was planned to create one reservoir with a capacity of 3km^3 and an expanse of more that 150 km^2.

In Azerbaijan, several hydroelectric rehabilitation projects are in progress. When completed, these projects should result in an additional 671 megawatts of electricity capacity. The 360-megawatt Mingechaur hydroelectric project on the Kura River is estimated to cost $41 million and is scheduled for completion in 2001. The EBRD loaned the country $21 million to finance the replacement of generators at the plant, as well as to install environmental controls. The Islamic Development Bank and the European Union's Tacis City Twinning programme are cosponsoring the effort. Plans are also being discussed by state power company Axerenerji for a $42.5 million development of small hydroelectric stations in the autonomous Nakhichevan region. The most promising scheme involves construction of a 23.1 megawatt capacity four-station cascade on the Gilan river. Other proposed projects include a $9.8 million, 4.5-megawatt hydropower station on the Vaykhyr river and an $85 million, 31.5-megawatt plant on the Ordubad River. The Islamic Development Bank has expressed interest in assisting with the projects. Azerbaijan had considered potential wind power development for Nakhichevan, but initial studies showed that every wind-generated kilowatthour would cost twice as much as a hydro-generated kilowatthour.

Recently the World Bank has supported the Armenia Dam Safety Project essentially because the "failures/number of existing dams" ratio is increasing, and the ratio is considerably higher than the value of 0.5% for dams built after 1951 (ICOLD, 1995), In all cases, proper surveillance could have detected the incipient distress in time to avoid failure. No dam was reported to have functioning instrumentation, and country-wide around 460,000 people are at risk. In light of this situation, the primary objective of the project is to protect the population and the socioeconomic infrastructure down-stream of the dams facing the highest risk of failure. Achieving this will also enhance the sustainability of dam-based economic activities in Armenia. The Dam Maintenance Enterprise (DME) will be responsible for the safety of the dams being rehabilitated under the project.

Slovakia

The Slovak water management industry has consistently relied upon a purely technocratic approach to addressing water-management problems (Kravcik et al. 2002). This approach strongly stems from the former communist regime's social perspective. During the communist administration, over 60 large dams were constructed, which resulted in the destruction of over 80 historical villages and the displacement of more than 60,000 citizens. On the majority of large rivers, flood-prevention dams were built, and this resulted in the drainage over 500,000 hectares of farmland and the canalisation of over 8,000 kilometre of streams. Completed after the communist regime's fall, the Gabčíkovo dam, found on the Danube River, has become a symbol of the technocratic paradigm in Slovakia.

Slovak water management leaders have, since 1989 regime changes, started to build additional dams from the state funds (i.e. from publics'/taxpayer's money). This is how they managed to build yet another dam at the Váh river. The Žilina dam project represents another public expenditure with the planned costs topping US$ 100 million. While the dam is not yet completed, its costs now stand at over US$ 250 million. A further example of the costs imposed by this technocratic approach is the Turèek dam. This addition project destroyed a four-hundred-year old medieval monument through its construction of the Turèek water-supply system, which maintains an average capacity 460 litre per second. A further proposed dam project was the Tichý Potok dam, discussed below, which would entail a 65 metre wall and 700 litre per second capacity. It was planned as a water-supply resource for Eastern Slovakia cities, and its proposed costs were to amount to US$ 200 million in 1994.

The pressure of the concrete lobby to build new dams in Slovakia is enormous, as developments at Gabcikovo, Zilina, Turcek, Slatina and Tichy Potok clearly demonstrates (Biodiversity in Slovakia, 1998). There are currently active plans for new reservoirs at Slatinka, Turcek, Tichy Potok and elsewhere, though the last of these was shelved after intense opposition from Slovakian environmentalists. The latter proposal could result in the flooding inundation of five historical villages (Tichy Potok, Olsavica, Nizne and Vysne Repase, and Torysky) (People and Water, 2002; REC Bulletin #72; *World Rivers Review*, 12(3), June 1997).

The story began in 1954, when the Czechoslovakian state water company, spurred by the lack of adequate drinking water in eastern Slovakia, proposed construction of a 65-meter dam on the Torysa River, creating a reservoir just west of the small village of Tichy Potok. Over the next 25 years, many studies were undertaken for the project, but no construction was ever begun. In the 1990s the project was dusted off and re-presented to Ministers, who were only too glad to consider it. Still, major damage had already been done to six villages near the proposed site. The water company's intention to build the dam at Tichy Potok had originally meant the evacuation of these villages, many of them over 700 years old. Living under the threat of forced evacuation and legally unable to build or expand anything in their villages for over 30 years, the villages were nearly killed by attrition, population dropping in the period from 6,000 to 1,500 inhabitants.

Slovenia

Slovenia is also working to upgrade its hydroelectric facilities. The Slovenian company Dravska Elektrarne (DEM) is in the process of upgrading the Vuhred and Ozbalt hydroelectric plants. This will be the second set of hydroelectric repairs and upgrades since DEM began the programme in 1993. The first stage, which is nearly complete, involved raising the installed capacity of the Mariborski Otok, Dravograd, and Vuzenica hydroprojects by a combined 34 megawatts. Installed capacity at Vuhred and Ozbalt will be increased by 31 and 39 megawatts, respectively.

Yugoslavia

There are plans to rehabilitate the 2,100-megawatt Iron Gates 1 hydroelectric plant, and Serbian utility EPS announced a tender to appoint a partner in the effort in June 2000. The plant is jointly owned and operated by Romania and Serbia, which plans to restore its half of the 12 turbine generators on the project. The upgrade will increase the installed capacity of each of the turbines by 15 megawatts (a 10% increase). The project will cost an estimated $100 million and will be undertaken between 2001 and 2008. The Romanians are in the midst of restoring and upgrading their six generators at the plant.

Concluding Comments: Towards a Political Ecology of Dams and Development

World Bank involvement with dam-building has been in decline over the last two decades. Between 1970 and 1985, the World Bank was completing about 26 dam related projects per year, eight of which involved direct financing of the dam while the rest involved funding associated facilities. From 1985 to 1995, the World Bank approved 39 projects that included construction of a new dam, or roughly four per year, which is about half the rate from 1970 to 1985. This downward activity trend has been continuing with an average of roughly three projects with new dams being approved per year over the last four years. Currently, the World Bank has six new dam projects in the pipeline, which may be approved within the next two years, so this trend of lower activity will continue in the near term. Notwithstanding this good news, the collapse of the centrally-planned economies of Eastern Europe in 1989 and their subsequent long path to integration within the EU has created a golden opportunity for European dam boosters to foist a new phase of development on the weakened states of the region. As Lecornu (1999) points out:

These changes are opening up excellent opportunities for high-quality energy sources, especially when they can offer firm capacity. We should therefore see two developments, especially in Europe. Firstly, greater interest in hydro dams controlling a storage reservoir, and secondly, gradual extension of existing hydro-electric generating plant, and even more pumped storage schemes.

In this chapter we have explored the long history and persisting allure of hydrological engineering projects; as solutions to a variety of practical problems, as mechanisms for regional re-integration, as celebrations of national power, etc. As we shall see in the next chapter, the sobering realities of climate change too have been drafted into the 21st century dam-makers' armoury of arguments. Whilst the author is not implacably "anti-dam", it does seem to me that the arguments for more dams in Europe are generally more about the self-interest of certain key stakeholders than they are about the common good. Consequently it behoves us to begin to think about such proposals as complex articulations of sociopolitical and material needs, in other words the need to protect our communities from floods or provide drinking water supplies may be materially undeniable, but *how* we elect to understand these needs and then meet them is clearly political. As Greenberg and Park put it in the inaugural issue of the *Journal of Political Ecology* back in 1994:

> Political ecology does not amount to a new programme for intellectual deforestation, rather it is an historical outgrowth of the central questions asked by the social sciences about the relations between human society, viewed in its biocultural political complexity, and a significantly humanised nature.

Suggested Activities for Further Learning:

1. Though this is primarily a book about managing freshwater resources, some of the largest dam structures currently under contemplation are "barrage" across estuaries, designed to generate power. On of the most ambitious of these is the proposed Severn Barrage, which could generate something like 10% of total electricity demand in England and Wales. Research barrages, and perhaps the Severn Barrage in particular, and prepare a list of pros and cons.

2. In British Columbia Canada, provincial officials are actively promoting so-called "Independent Power Producers" deals (IPPs) whereby run-of-river hydroelectric power plants are installed on the province's rivers in return for a concession fee and with all the electricity generated sold into the publically-owned power company, the aptly named BC Hydro. Prepare a case study of one or more of these IPPs.

3. The Karnajakhur dam project in northeastern Iceland has proven extremely divisive politically, but what are its real costs and benefits? Prepare a cost-benefit statement for this project based on the best available data (see also the next chapter).

Chapter 8

Climate Change, Social Change and Europe's Water Futures

Introduction

As an English farmer once said to me: "The problem with water is that there is either too much of it or too little of it. Either way we're screwed." The irony is that these words were spoken in 2004 and were meant as hyperbole, not prophesy. At that time the UK was experiencing a prolonged drought that was only really broken by the catastrophic floods of June and July 2007; floods that triggered the largest peacetime mobilisation of civil defence activities in UK history. Somewhat counter-intuitively (given recent weather!), the UK government is now suggesting that climate change means that we in the UK will have to learn how to get by with *less* water – as low as 120 litres/person/day (approx 32 gallons/person/day) according to the 2008 report *Future Water: the government's water strategy for England*. To put this into perspective – the World Health Organisation argues that the public health-related floor on water consumption is somewhere around 50 litres person/day, even the most parsimonious European country has consumption rates around 120 litres/person/day and – at the other end of the consumptive spectrum – US water consumption is closer to 400 litres/person/day (even in the desert southwest!). Still, whilst reservoirs in northwestern Europe are currently full, other parts of Europe, particularly in the Mediterranean Basin, are nearly tapped out (EEA, 2009).

As with my farmer friend, according to the mass media sometimes it's flood, sometimes it's drought, but water is *always* a problem. But what kind of a "problem" is water, exactly? Though we readily agree that it is a "problem", unpacking the exact nature of that problem has proven difficult and indeed contrary positions are taken by key stakeholders. This penultimate chapter to this book I discuss the dimensions of the water "problem" under the following subheadings:

1. Hydrological Implications of Climate Change
2. Water as Resource
3. Water as Hazard
4. A Regional Geography of Europe's Water Challenges
5. Water and Sustainability in the 21st Century

Through reframing our conceptualisation of water as both a resource and a hazard with "socio-natural" imbrications (one of the leitmotifs of this volume) I hope we can move towards a critical platform for thinking productively about the potential of that chameleon-concept "sustainability" in its application to water resource management. In the preceding seven chapters we have built up a conceptualisation of water as something which is constituted out of natural (hydrology, biogeography, chemical, etc.) and social (economic, political, cultural, etc.) processes – it is a "socionatural" good par excellence. Consequently the reader should not be surprised that I take a similar approach to the question of climate change's likely impacts on European water management. After briefly introducing the Janus-twins of too little water and too much water I turn to a consideration of how they are manifesting themselves in four broadly conceived European regions. I do not provide more than a brief discussion of issues in western, northern, eastern and southern Europe, but I do claim to have "Europeanised" the topics examined in this book by so doing. In the final section of the chapter I consider the key challenges to more sustainable water management and in particular the extent to which "integrated water resource management" and the EU "Water Framework Directive" can potentially take us there.

Hydrological Implications of Climate Change

We are by now accustomed to reading that climate change is reducing the amount of water that is available to even traditionally well-watered parts of Europe, such as Britain, Germany and Poland. Eleven of the twelve years in the period (1995-2006) rank among the top 12 warmest years in the instrumental record (dating back to about 1850). According to the IPPC 4th Assessment, released in late 2007, we have seen a 0.74 °C increase in global average temperature – 20% greater than previously thought (IPCC 2007). While there is much debate about the precise nature of the interactions between key factors such as global CO_2 concentrations, the Jet Stream and the Atlantic Thermohaline Circulation there is no longer any doubt that we live in an era of warming. Moreover, this warming, as Sir Nicholas Stern pointed out so forcefully in his ground-breaking report, threatens the basis of human existence in many world regions, particularly in the middle latitudes.

Of course it is not just that certain European regions may have to learn to do with less water in absolute terms, although some European countries in the Mediterranean Basin are currently planning for a 30-40% reduction in incident precipitation by mid-century. Indeed the Fourth IPCC report, *Climate Change 2007* declared that dry regions are projected to get drier, and wet regions are projected to get wetter:

> By mid-century, annual average river runoff and water availability are projected to increase by 10-40% at high latitudes and in some wet tropical areas, and decrease by 10-30% over some dry regions at mid-latitudes and in the dry tropics...

Climate change also means that we are likely to experience an increase in the intensity, severity and frequency of extreme weather events such as droughts, storms and floods. Further, regions at risk of drought are likely to become larger, taking in much of southern Europe, the Mediterranean Basin and even swathes of central Europe (EEA, 2009; Lloyd-Hughes and Saunders, 2002; Bradford, 2000). Differences in water availability between regions will become increasingly pronounced. Areas that are already relatively dry, such as the Mediterranean basin and parts of Southern Africa and South America, are likely to experience further decreases in water availability, for example several (but not all) climate models predict up to a 30% decrease in annual runoff in these regions for a 2°C global temperature rise and 40-50% for 4°C. In contrast, parts of Northern Europe and Russia are likely to experience increases in water availability (runoff), for example a 10-20% increase for a 2°C temperature rise and slightly greater increases for 4°C, according to several climate models. In all cases however existing drought conditions will intensify, particularly as regards those regions affected by winter droughts (Central Eastern Europe, Western Russia, Northwestern Europe including Norway).

Even more worryingly – and often missed – is the fact that water supplies stored in glaciers and snow cover will be reduced over the course of the century – glaciers are receding at alarming rates across the world and European glaciers offer no exception to this trend. According to Matthias Huss of the Zurich university's Laboratory of Hydraulics, Hydrology and Glaciology "Glaciers are starting to lose mass increasingly fast", with serious implications for freshwater availability in many central European countries (BBC, 15 December 2008). The impact of glacial recession on water availability will vary with the specific region or country under the microscope (as we shall see Norway may be particularly vulnerable). Less easy to project is the effect on regional climate of the complete or near-complete removal of glaciers and snow cover – this is the subject of much current research throughout Alpine Europe. The basic issue is that we may not be able to rely on seasonal release from snowpacks and glaciers to augment river flows through summer months, exacerbating whatever deficits may also be occasioned by lower precipitation. Unfortunately for us all of this implies that there may be a mutually reinforcing dynamic between the warming trend and the drying trend.

Simultaneously heavy precipitation events are very likely to become more common and will increase flood risk. Recent regional climate change modelling by the UK-based Hadley Centre (HAD RM3H) shows that the likelihood of intense rainfall events – such as the historic flooding in England in June and July 2007 – also increases. Ironically, heavy rain can also contribute to water shortages. Extreme droughts and precipitation events cause adverse effects on water environments, creating not only a water shortage/surplus problem but a larger environmental issue as well. Since the water-holding capacity of air increases exponentially with temperature, the water cycle will intensify, leading to *more* severe floods and droughts. There will also be more energy to drive storms and hurricanes (Stern Review, 2007; Lloyd-Hughes and Saunders, 2002; Bradford, 2000).

Some find it hard to believe that flood and drought could actually come hand in hand, but this is the current reality. Indeed, as pointed out in Chapter 7, one reason for the proposal of new dams and reservoirs is simply that alterations in the hydrological regime will require adaptations to the hydro-technical infrastructures. Put more simply – British reservoirs which were originally developed such that they could be topped up with relatively regular and modest rainfall patterns (as prevailed at the time of their creation) must now be re-engineered to make the most of periods of increasingly intensive precipitation separated by periods of relative drought.

All of this is taking place at the same time that human consumption of water is increasing significantly – absolute consumption is likely to double between 2000 and 2025 to as much as 5500 km^3/yr or about 50% of available water resources. Impacts of water shortage will be concentrated not just in regions of relative scarcity but in regions of particularly intensive use – such as our burgeoning megacities more than two dozen of which may be classed as "water stressed" by 2015 (Jenerette and Larsen, 2006; Niemczynowicz, 1996).

So, contra the Candidean prognostications of ill-informed pundits such as Bjorn Lomborg, the water crisis is here, it is real and it is going to intensify whatever we do now and in the future. But the altered hydrological regime will have other, less obvious, impacts too. The discharge of treated wastewater effluent into rivers with relatively low flows means that less dilution (a key mechanism for purifying water – see Chapter 4) takes place than would under normal conditions. Water quality and availability are closely linked, so lower quality also means there is less water available for abstraction. Moreover, intense rainfall results in surface flooding rather than infiltration to underground aquifers. Water quality problems arise through surface water run-off carrying pollutants from land and transferring them into rivers and lakes. While such pollution occurs under normal conditions, it is intensified during storms and flooding events.

The Stern Review Report, published in late 2007 took an economistic perspective to the IPCC prognostications and made a series of startling predictions about the likely impacts on water availability and their socio-economic ramifications. It predicts, for example, that even a mid-range temperature rise of 3°C could mean that southern Europe experiences serious droughts once a decade. In fact, given the droughts of 1976, 1995 and 2003 this reality may already be here. Stern also predicts that with this modest temperate rise many agricultural regions of middle latitudes will become far less productive and that of higher latitudes will peak. And the environmental devastation could be severe with species losses of:

- 20-60% of mammals
- 30-40% of birds
- 15-70% of butterflies

And if these alterations compromise global circulation systems like the Atlantic Thermohaline Circulation or the Jet Stream the cumulative effects could be even more severe. Many scientists contend that it is the potential mutual reinforcement of distinct processes such as the CO_2 concentrations, sea water temperature, the Jet Stream and the Atlantic Thermohaline circulation that could lead to quite abrupt, as opposed to gradual, climate change (National Academies, 2008).

This is not the end of the matter however for although the climate science may well be overwhelming and unequivocal, public sentiments have proven harder to shift. Very recently I had a "robust" discussion with a local publican (I live in the English county of Gloucestershire, near the epicentre of the summer 2007 floods) about whether or not individuals could or even ought to do anything to combat climate change. The publican's central points were twofold: on the one hand denying (without any real counterevidence) that significant anthropogenic climate change was taking place and on the other claiming that individuals could not really do much to alleviate or mitigate even if it were occurring. The former argument is well-known, given that even the former president of the United States seemed to think that his opinion somehow outweighs the accumulated scientific work of over 1000 scientists in 150 countries.[1] Usually I deal with this argument by asking the respondent which part of the detailed and lengthy IPCC, Stern and other assessments they care to differ with – they invariably find that they are ill-equipped to do so! After all the science of heat-salt exchanges in oceanic bodies is complex stuff and getting more so all the time. But the point surely is that there is much to do in communicating effectively the realities of climate change to a wider audience. The second part of my publican-friend's objections will be familiar to anyone who has read their Garret Hardin[2] – what can individuals possibly do if other players are steadfastly refusing to conserve? Here I generally appeal to two arguments:

1. An inter-generational perspective. This works especially well for anyone who has children – the counter is to ask if there is anything they would *not* do to safeguard their children's futures? *Even otherwise ardent neoconservative Margaret Thatcher consistently argued that our environment was on loan from future generations and was not ours to despoil.*

2. A thought experiment inspired by the anthropologist and biogeographer Jared Diamond. In *Collapse* Diamond asks readers to consider what the impending collapse of Mayan, Easter Island, Greenland Viking and other societies might have looked like *before* and *as* it happened, with all the structural similarities of mounting evidence but uncertainty in prediction?

---

1   Moreover, sensationalist documentaries like "The Great Climate Change Swindle" try to make big profits (and reputations) by slandering and misquoting senior scientists – thankfully in July 2008 the makers of that documentary were forced by UK regulatory authorities to apologise for their misrepresentations.

2   In 1968 Garrett Hardin published an article "The Tragedy of the Commons" in the journal *Science* which seemed to suggest that collective common resource management must necessarily fail.

As Diamond points out, ignoring alarm calls just because the last few may have been false or because there remains some level of scientific uncertainty, risks a cataclysm.

All these lines of argument lead me to what I sometimes call "small "c" conservatism" (actually it is probably better called "pragmatism") – the idea embedded in European (and I daresay North American) culture that waste is morally wrong and thrift is morally right. There are social, political, economic and cultural dimensions to this sort of conservatism or pragmatism, but such as position leads us to consider hard trade-offs as we shall see: for example, does the individuation and pressurisation of water consumers attendant on compulsory water metering (now under consideration in England and Wales) offer benefits that can offset the likely erosion of social equity in a key natural resource? Will we need to consider moving mass populations from seriously impacted regions to less impacted regions, and perhaps even abandoning entire mid-latitude nations as they cease to be able to support human life? Alternatively, how far are we willing to go to mitigate some of the effects of climate change – might we reconsider sovereigntist attitudes towards bulk water exports?

Water as Resource

Talk of regional water shortages in the UK often triggers questions that are essentially about the creation of new supply. Why can we not build new reservoirs or desalination plants? Why not build a national water grid? Or, more controversially, why not expand our ability to treat "blackwater" (water carrying human waste) so that it can be reused – the so-called "toilet to tap" solution.

The idea of a national water grid which would redistribute water around the nation like the Great Central Water Carrier in Israel, the National Water Plan in Spain or the Central Arizona Project in Arizona is periodically raised and discussed. However, there are many reasons why such a project is unlikely to be seriously considered in the UK, thus throwing the emphasis back on local supply and conservation measures.

First, unlike gas and electricity, *water is heavy*. Indeed, the average European family of four uses water weighing about two-thirds of a tonne each day. Transporting it over large distances would require major engineering work and pumping operations that would consume large amounts of expensive energy. The water industry is already fairly energy intensive and consumes about 3% of total energy used in the UK. Most of this is used to pump input water and wastewater and to run treatment plants that ensure our water meets strict environmental and health quality standards. The industry is responsible for approximately four million tonnes of greenhouse gas emission (CO_2 equivalent) every year. That's less than 1% of total UK emissions but is rising gradually year on year. On the one hand we're getting more efficient at abstracting, treating and supplying water and wastewater services and on the other hand population and consumption growth, along with more stringent standards, are driving energy use up. Carbon reduction is already a major challenge facing the industry.

Second, the possibility of a national grid is also questioned on economic grounds. Since water is relatively cheap (around £0.15/litre as opposed to more than £1.00/litre for petrol) suggests that a national grid would be disproportionately expensive. Building a national water grid to ensure the provision of amounts above and beyond this level would be outrageously expensive, even when compared with alternatives like desalination.

Of course any such plan would attract considerable environmental criticism. Transferring large volumes of water away from an area is likely to cause big changes in the local ecology, changes that many would see as damaging to flora and fauna. The transit route would be significantly affected, not least by the major groundworks required (even assuming no significant impact on transit communities). Moreover since there is a relationship between the finer details of water quality and local ecologies around the country, importing "alien" water could begin to alter the ecologies of the recipient regions.

For all these reasons, neither the government nor the industry is not seriously considering a national water grid. In a very real sense the time for these sorts of projects has come and gone – there is no longer a socio-natural consensus that would support such megaprojects as the Hungarian/Slovak, Spanish, Israeli or Turkish ones. In fact the heated contest between Hungary and Slovakia in the 1980s over Gabcikovo–Nagymaros (discussed in Chapter 3) is indicative that even in post-communist countries (where the policy network is presumably still more inclined towards the former "Engineering Paradigm") there is waning support.

As noted in earlier chapters such supply-side solutions to water shortage are currently out of favour, with policy-makers preferring to focus their energies of reducing demand across the entire range of users. Thus the government's strategy for water provision, *Future Water*, talks about revising architectural and planning codes to encourage water conservation. Whilst rainwater harvesting and greywater reuse are technically possible, they tend not be cost effective when applied at the level of individual dwellings or retrofitted to existing housing stock. Across Europe the situation is broadly the same – though some governments are pursuing completion of national water grids long under development (e.g. Spain), most are now concentrating their energies of disciplining demand. And, as we have already seen, demand management is occurring largely through economic mechanisms related to the redefinition of water services, in the 1980s and 1990s, as "a scarce resource [that] has an economic value in all its competing uses" to use the language of the 1992 Dublin Statement.

Overall we are left with a number of key questions to which policymakers have as yet provided no answers. For example, questions about metering are intrinsically linked to questions about pricing; therefore, metering discussions should not be taken as separate or an add-on. If one of the purposes of metering is to discipline growing demand for water then we need to consider very carefully the experience in other European countries and in North America where metering is already much more widespread than it is in the UK and the case for a conservation effect is not proven. Moreover the relations between drinking water versus toilet use, outside use, sewerage and site area drainage portions of metered (and unmetered) bills need to be carefully considered.[3] In a nutshell there needs to be a clear and thoroughgoing debate regarding the precise purpose(s) of metering and the links between different types of metering and different types of desired outcomes. As Medd (2005; 2007) observes this will require a careful rethinking of the purposes modern European culture adduces to water and perhaps a differentiation between different sorts of water provision/use (e.g. greater encouragement of rainwater collection and use, grey water reuse, etc.). In the UK case it is clear to see that whilst existing stakeholders all talk about metering, their perceptions of and expectations for meters are divergent. Thus it is incumbent upon regulatory authorities to create a strong, well-supported and *unified* voice for smart metering options which can achieve environmental, economic and social equity goals. Moreover the argument about the purported fairness of water meters is widely recognised as just another version of the argument for "user fees" which, in extremis, erodes the social basis of modern political society and may represent yet another form of "governmentalisation by stealth" of daily life (Bakker, 2001; 2005; Jenkins, 2006).

Water as Hazard

Both too much water and too little can create natural hazards and, as Jared Diamond, Joseph Tainter and other anthropologists of past societal collapses remind us, can contribute to social and political unrest. The basic problem is that human societies have not proven particularly adept at foreseeing fundamental challenges to their viability. Too often a combination of (fallacious) reasoning by analogy with past challenges our human tendency to "normalise" even abnormal events if we are exposed to them for any length of time and also the "landscape amnesia" effect whereby we tend to forget even important things like the distribution of rain events around the year or the preponderance of certain species of plant or animal (which can act like a canary in a cage). All of these factors add up, for Diamond, to the fact that human groups can often do bizarrely irrational things even if most of their constituent members are otherwise rational and reasonable. The flooding in Gloucestershire England in July 2007 provides many expressions of this essential paradox.

---

3 Consumers tend to think only about drinking water and not about sewerage, wastewater and drainage as part of the metering issue. In England and Wales metering advocates frequently overlook the fact that the key reason why metered bills differ from unmetered bills is because they are strongly administered – thus the comparison is unsound.

In summer 2007 Gloucestershire experienced the largest water-related civil defence emergency since World War II. The severe flooding which affected much of the country during June and July 2007 followed the wettest-ever May to July period since national records began in 1766. Met Office records show that an average of 414.1mm of rain fell across England and Wales – well over double usual levels. In Gloucestershire total July rainfall of 414 mm was 4 times the 30 year average. The additional volume of rain which fell from May to July was 31,140 million cubic metres – more than four times the amount of water in all the lakes in England and Wales combined. The sheer volume and intensity of the water overwhelmed many drainage systems and a number of defences (Pitt Review, 2007).

Over 6,000 properties were affected by the July floods, many of which were first flooded by surface water or by watercourses which reacted quickly to local run-off. The same properties were then flooded by the River Severn a few days later. Roads and transport links were affected by the floods and seriously hampered

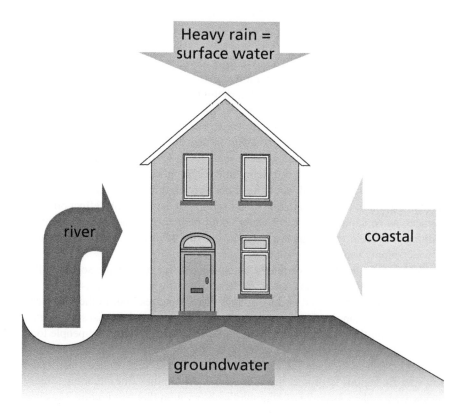

Figure 8.1   Types of flooding
Source: Author.

people's travel plans. Over 30 schools were damaged. Some 10,000 people were left stranded on the M5 motorway and had to be serviced and/or rescued by the Army. The Gloucestershire fire and rescue services attended 1,800 calls within 12 hours, almost 25% of their annual call volume (Gloucestershire County Council 2007). 420,000 houses were left without drinking water and 50,000 homes were without electricity.

The intensity of this event was compounded by the fact that it combined three of the four basic types of flooding, fluvial, pluvial and groundwater. In the first instance, heavy rains caused rapid pluvial (from the Latin for rain – "pluvius") flooding. Because of the surface hydrology of the region, the dumping of intensive rain on the upstream sections of the Severn River catchment system meant that within hours the middle and lower sections of the river overtopped and caused a second "wave" of fluvial (from the Latin for river "fluvius") flooding. This was then followed by flooding from the supersaturated groundwaters themselves. As one witness (cited in Thomas and Wilson, 2007) put it:

> We were sat at the end of our drive watching...and all of a sudden it came up
> through the grate...10 seconds later Jane's house were gone [sic].

All of this was caused, ultimately, by a flow of warm wet air from the southeast – something quite unusual for the region at that time of year.

The floods had a serious impact on agriculture, in England 1.3 million hectares of agricultural land is situated on a floodplain (Environment Agency, 2007). During the floods many crops such as wheat, barley and potatoes were damaged. This damage to crops is not only costly to farmer in the short run but contributed to significant price rises in the grocery markets in following months. After the rainfall had stopped it took a very long time for the flood water to clear, once the water drained away the land would have been covered in dirty residue left by the dirty flood water. Risk of contamination had a particularly serious effect on dairy farmers, as contaminated feed or water could permanently damage the animals' milking productivity. Milking cows must drink 70-100 litres of water a day, so loss of water supply was a huge challenge.

To its credit the UK government launched a major review, the so-called Pitt Review, of the flooding, its causes, the emergency response and lessons for the future. Its report, published in 2008 makes fascinating reading. In the first place it appears that many residents were only vaguely aware of the risks associated with living on the largest floodplain in the UK! Bizarrely, too many people took the view that government or the water company should deal with "the problem" and also compensate them for any losses. Even worse, few residents had taken any specific actions to protect themselves or their families from flooding. Surely this is an excellent example of Diamond's observation that "groupthink" is often irrational. Intriguingly the Pitt Review adds that:

In the hours and minutes before the flooding took place, there was a polarisation in awareness of the risk of flooding. Farmers and rural businesses were more likely to monitor the state of the weather and of the land around them, and were most likely to have been aware of the risk. Many others, often householders, were less aware. There were mixed experiences in relation to warnings. Some received flood warnings but disregarded these. Others sought information, and were reassured that there was no risk of flooding. Others did not seek or receive warnings, and remained unaware of the risk. Few, except farmers and some businesses, took any action...*For many, seeing the water coming into their homes or businesses was the first awareness of the real risk of flooding* (Pitt, 2007, my italics).

This appears to be human nature – an in-built tendency to accentuate positives and to downplay negatives. One outcome is that individuals frequently choose to inhabit localities that are, objectively, "risky".

Another is that our collective organisations – government departments, regulatory authorities, etc. – can be remarkable obtuse about truly factoring drought or flood risks into their development control systems. Risks to the built environment and human life are likely to be most acute in built-up coastal areas, which will be subject to all four types of flooding indicated earlier, and which are the product of decades of lax building control. Weston-super-Mare, located on the Bristol Channel in southwestern England, is one such location, which recent modelling has shown to be highly vulnerable to even 1 in 10 surge events (Thomas, Ingrams, Dhayatker, Searle, 2008).And it certainly does not help matters that the adjacent Bristol Channel has some of the highest tides in the world. The key question for English planning authorities is how can development control and the planning process more generally be used to add what Sir Michael Pitt calls "resilience" to increasingly vulnerable urban areas.

Now, I could have chosen to develop any of a number of other examples, including the flooding on the Odra River in 1997 or the heatwave of summer 2003 during which several thousands of people died in Western Europe, but the essential point is, I think, made. Water is a hazard (or a resource) not merely because of its physical presence or absence (or indeed its *sudden* presence!), but also and essentially because of the ways in which human societies have adapted to it. Such adaptations are in fact the sedimented results of experiences of the past and expectations about the future and about how others will react. One of our most powerful collective actors is of course the EU. Martin Geiger, the head of the Freshwater Program at environmental group WWF, has suggested that the European Union needs to wake up to how precious a resource water is: "If the EU would link water savings to the subsidies they provide, this would impact the current water use because there are efficiently gains of up to 30 to 40% you can reach with modern technology, and only a few areas currently use those modern technologies" (Deustche Welle, 2006).

A Regional Perspective on Europe's Water Challenges

Much about the likely hydrosocial futures for European countries has already been noted in previous chapters. We know, for example, that the pronounced shift towards demand management and Catchment Basin level management represents not the collapse of the "engineering paradigm" exactly, but rather its revitalisation through new technological fixes, in particular water saving technologies and water metering and management systems such as Integrated Water Resources Management as operationalised through the WFD. And whilst most Western European countries have faced similar economic, political and climatic challenges in recent years, their different initial conditions mean that their experience has been diverse. Here we have space to make only brief comments about recent water management issues in four broadly drawn regions: Western, Northern, Eastern and Southern Europe.

Western Europe

Water rights in France are owned by local communes and most supply and sewerage is outsourced to private corporations (in most cases Veolia, Saur or Suez) that maintain the water networks, provide billing, and develop infrastructure in consultation with the commune. In recent years persistent drought has forced the *Ministère de l'écologie, du développement et de l'aménagement durables*, in consultation with the regional *Agences de l'eau* to impose restrictions. During the pronounced drought of 2003 more than 14,000 people died from heat and dehydration-related causes as temperatures soared to well over 40 Celsius for much of August (Bhattacharya, 2003). Blame for the high death toll was placed on overstretched emergency services and on the apparent abandonment of so many elderly people. Five years later, and after 18 months of higher than average rainfall, many parts of France still suffer from water deficits and the regional to local level of regulation.

Unfortunately for French consumers there is little competition and tariffs are at least 20% higher than they otherwise might have been (Carpentier et al., 2004) though they are still lower than the UK or Germany. Leakage rates are among the highest in Europe at 25-30%. The 2006 *Law on Aquatic Environments* seeks to harmonise French water law with the WFD (Bongaerts, 2002).

In Germany water management is a joint federal-lander responsibility, with the federal authorities responsible for establishing only a framework for management and national standards. Lander governments are therefore primary competent authorities and have actively invested in the creation of water-technical and water-administrative infrastructures which are quickly establishing international reputations as centres of innovation (Kompetenzzentrum, 2008).

German domestic water use has declined from 140 l/p/day to about 120 l/p/d and consumer satisfaction is very high. Most drinking water is provided by the many small municipal companies although there are a few large private companies

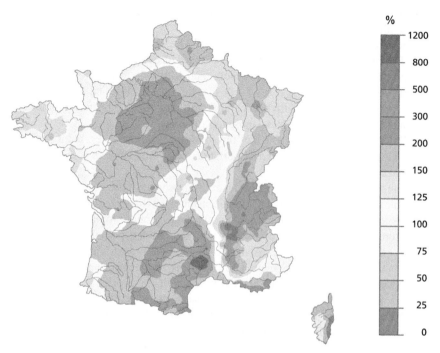

%

- 1200
- 800
- 500
- 300
- 200
- 150
- 125
- 100
- 75
- 50
- 25
- 0

Figure 8.2   Areas of water deficit and surplus

Source: Ministère de l'écologie de la France, "Bulletin de situation hydrologique".

like Berliner Wasserbetriebe and Veolia (OECD, 2004). Domestic water prices are the highest in Europe. Even so, continuing urbanisation – especially in the Berlin urban region – and climate change are combined to place great pressure on water supplies. Germany is also suffering from the effects of climate change with steadily rising temperatures and increasingly variable rainfall. Already agricultural production has been affected and some rivers are seeing instances of desiccation-related fish kills (Deustche Welle, 2006). Though localised water supply problems have been addressed through inter-regional water transfers and groundwater pumping, there are signs that these may not be long-term solutions. The action programme for the Rhine River, Germany's most important waterway, has set quality targets which may prove ever harder to reach and to date only about 60% of surface waters have reached the quality standards set by law.

For these reasons Germany has been at or near the forefront in the development of new water technologies such as rainwater harvesting and grey-water reuse. Both systems have been integrated into the German spatial planning and building specification framework, leading to exemplar developments such as the Football World Cup Stadium in Berlin which incorporates more than 10,000 m³ of storage for rainwater harvested from the roof structure. Rainwater collection and circulation systems are also used for cooling applications, in place of energy-intensive air conditioning systems (Koenig, 2007).

The relatively low-lying alluvial lands of The Netherlands are drained, and indeed were largely created, by the lower reaches of one of Europe's great rivers: the Rhine (ten Brinke, 2005). Rising more than 1,200 km away in the mountains of Switzerland the Rhine is a relatively fast flowing river which upon entering The Netherlands splits into myriad natural and human-made channels, branches and canals – creating the water landscape for which the country is so famous. The Rhine not only provides much of the country's water supplies, but it is also a primary artery of commerce and transport and, inevitably, a source of ever-present flood hazard.

In the Netherlands water supply problems are a product of the unusual "randstad" economic geography of the country, climate change and changing social perceptions and pressures. With all urban areas in this compact nation quite proximate to intensive agriculture there is an endemic problem with agricultural pollution, particularly nitrates, nitrates and phosphates. Also, the predominantly low-lying geography of the nation (nothing over 400 metres ASL) means that water planners are at the forefront of planning for climate change in terms of flood defense – the abiding rule now being "allowing water to find its place". Van der Brugge and Rotmans (2007) are also concerned that "water" has become a "persistent" problematic in Dutch popular discourse, a problem which itself requires careful management. Whilst current trends in water management are in many ways centralising, the Dutch tradition in water planning is highly decentralised and Ravestiejn and Kroesen (2007) contend that this places strain on Dutch water policy culture. Dutch governmental policy and popular sentiment see water as a public service and, unlike in other parts of Europe, the trend has been towards strengthening public control over water supply utilities (OECD, 2004). Further pressure is created by climate change which is both drying and rendering more volatile the European climate. In recent years the Dutch Meterological Service has been registering the hottest and driest summers since record-keeping started in 1705.

Northern Europe

In Norway water issues take a very different shape. The country is not generally seen as water scarce, although as elsewhere water authorise are working to meet rising demands with a static or declining unit of resource. With the relentless contraction of the country's glaciers and the increasing uncertainty of the winter snowpack the nation's historic reliance on hydroelectricity is under threat. Norway is one of the top six nations in the world in terms of installed hydroelectric capacity and the number one in terms of per capita generation capacity. State-owned Norsk Hydro, founded in 1905 by engineer Sam Eyde in 1905, has been developing Norway's water power capacity for more than a century, particularly in the southern Telemark region. Virtually all of Norway's electricity needs are met through hydroelectricity generation.

**Photo 8.1   Sam Eyde with the first hydropower plant at Rjukan in back
ground**

Source: Author.

Notwithstanding Norway's relative abundance of surface water resources, droughts
in the mid-1990s led to catastrophic draw-downs on water supply reservoirs in
southern Norway particularly affecting the capital Oslo (Hagen, 2000). It has also
proven necessary to radically modernise the country's largely tacit system of water
regulation and even to consider strengthening the Scandinavian electricity grid to
ensure security of supply (Haugli, 1994). However, whilst the optimal trigger price
for investment in new hydropower may now have been reached (Kjaerland, 2007),
such investments have been rendered more uncertain due to climate change and
changing political priorities. In Simultaneously local municipal water suppliers
are having to look at new forms of finance, including public-private partnerships,
to pay for needed infrastructural maintenance and upgrading.

Perched on Europe's far northwestern perimeter tiny Iceland harbours a
surprising number of water problems. Though precipitation is expected, even
under the most pessimistic climate change projections, to remain strong, water
manages nevertheless to be a problem, primarily for three reasons.

First, the recession of glaciers worldwide has affected Europe's largest glacier,
the Vatnajokull, located in the southeast of the country. In addition to shrinkage of
the ice sheet, the region have become less stable geologically and as a result now

Photo 8.2   Godafoss, eastern Iceland

Source: Author.

suffers more risk from "jokulhaups"; sudden eruptions of under-ice volcanoes leading to intensive flooding events. The most recent major jokulhaup event took place in 1996, though there have been more minor eruptions in 1998 and 2004.

Second, in 2002 the Icelandic government approved the Karahnukar Hydro Power Project, to develop hydroelectric capacity of 800 MW with a series of dams and reservoirs of over 150 metres in height and covering more than 60 square kilometres in the remote eastern part of the country. This project is not at all about providing power for local people, but rather for an aluminium smelter that is intended to create local jobs. The irony is that not only does Iceland provide much of its own domestic electricity needs through hydroelectric and geothermal production, but it has no known bauxite reserves and so the aluminium production plant will be entirely dependent upon the fickle international market for aluminium.

Third Orkuvieta, the publically-owned multi-utility (including water supply) company for southwestern Iceland derives less than 10% of turnover from water supply, but nobody knows how the alterations of hydrology could affect the nation built on ice, liquid water and steam. Jonsdottir et al. (2005) used stochastic flow modelling to predict that there is a quarter of a chance of flow deficits which would adversely effect Iceland's hydropower system.

Eastern Europe

Poland is the 5th largest member-nation of the EU with a territory of just over 300,000 km² and a population of 38 millions. Geographically virtually all of its territory is situated on the Northern European Lowland meaning that the prevailing hydrological trend is for surface waters to drain northwards away from the Carpathian mountains which form its southern boundary with Ukraine, Slovakia and Czech Republic. Major river basins are formed around the Wisla (Vistula) and Odra (Oder) Rivers, along which are located the major population centres, including Warsaw, Krakow, Gdansk and Wroclaw. As we saw in Chapter 6, since 1989 there has been significant penetration into the Polish water sector by foreign multinationals, in particularly SAUR (Gdansk) and International Water (Poznan), although most water provision is still undertaken by municipal utilities. The Polish government has recently signalled its intention to accelerate privatisation of municipal water utilities through the creation of holding companies which may consolidate numerous unappealing small providers into more appealing larger providers (OECD, 2003). Interestingly, various case studies which have recently been published as part of the "Watertime" project have shown no correlation between privatisation and water utility efficiency – public water utilities (Lodz, Warsaw) are as likely to have seen improvements across a range of indicators as private utilities (Gdansk).

In common with Germany, currently European-level climate change projections suggest a general drying trend with increased volatility. Certainly flooding in the upper Oder basin in 1997 and in 2004 were triggered by abnormally heavy precipitation events, although their effects were magnified by decades of poor land management including deforestation, river channel modification and uncontrolled urban development. Ironically, such events can also exacerbate water supply shortages since local and regional impoundment structures are often ill-suited to capturing this type of rainfall for later use as a source of drinking water supply. At the same time Polish water managers must grapple with the relatively low quality of their water environment. Decades of centrally-planned industrial development has created a situation where few stretches of Poland's rivers reach the highest quality classification (Szulinska, 2002).

Bulgaria, a nation of 130,000 km² with a population of approximately 8 million became an EU member on 1 January 2007. Prior to the collapse of its last communist government in November of 1989 water services were supplied by the state company "Vodosnabdyavane I Kanalisatsiya". With the beginnings of state industrial dismantling in 1991 water regulation and water supply were administratively separated. Nominally Bulgaria's water resources are managed by the central government's National Council on Waters (NCW), whose members are appointed directly by the Cabinet of Ministers (Knight, et al., 1995; Staddon, 1995). At the same time the former ViKs were reorganised into 14 municipal 28 regional water companies. Not surprisingly the largest of these, the Sofiya ViK, became the target for takeover bids by foreign water multinationals, although by 1995 the process of water sector restructuring had descended into political farce (Staddon, 1996; 1998). This, combined with the pronounced European drought of the mid-1990s, created the preconditions for widespread supply failures affecting much of the capital region. Many districts of the city were reduced to receiving water only one day in three and queues for water became commonplace (Photo 8.3).

Acting in a state of emergency the caretaker government of Renata Indjova summarily approved the VK-Rila water diversion project and so set the stage for the next phase in Bulgaria's postcommunist hydro-politics.

On the morning of the 8 February 1995 troops of the Bulgarian Interior Ministry stormed barricades erected by local citizens angry about the VK-Rila project to divert water from their region to ameliorate the water crisis in the capital city. Against the motley opposition of farmers, unemployed miners, mothers with children, retirees, environmental activists and the not inconsiderable retinue of spectators marched the disciplined ranks of gray-uniformed national policemen carrying clubs, rifles and the weight of central government. Videotape of the event shows many elderly Bulgarian women pleading with the soldiers to turn back, to understand and empathise with their situation, even as the military phalanxes removed the barricades and brusquely pushed people back and out of the way of the oncoming construction trucks. Nineteen people were arrested and about a dozen were injured, including one person seriously injured in what the Interior Minister

Photo 8.3   People queuing at public water taps in central Sofiya, March 1994

later referred to as "a textbook model of police work guided by principles of democracy and respect for human rights" (*BTA Daily Bulletin*, 13 February 1995). Additionally, in a move reminiscent of Soviet-era practises, several journalists had their cameras confiscated and their film destroyed by police (Staddon, 1996). Construction of the diversion was completed in late March 1995.

Since that time both much and little have happened in Bulgaria's water sector. First, the drought of the mid-1990s was replaced by wetter, more hydrologically normal, times. This alleviated the immediate causes of the Djerman-Skakavitsa crisis, but did nothing for the structural causes. In late 1999 International Water Holdings purchased a 25 year concession to run Sofiiska Voda for profit, based the usual conditions designed to guarantee investment and keep prices down in return for certain financial returns to the company and assistance in leveraging investment funding from multilateral development banks (see Chapter 6). Unfortunately much of the promised investment has not been forthcoming and by 2004, with the return of drier than normal weather, the city was once again looking at water rationing regimes. Meanwhile water supply in the rest of the country remains largely unchanged with the second largest city, Plovdiv only receiving significant investment (from the EBRD) in its water sector in 2007.

Southern Europe

Spain is the EU's fourth largest country and has both a wealth of natural water resources and an extensively developed water infrastructure. More than a 1000 dams have been constructed and the National Water Plan calls for more (Swyngedouw, 1999). Desalination plants have also been extensively developed, particularly in the Canary Balearic Islands, providing for up to 220,000 m³/year. More than any other European country Spain has invested heavily in the creation of a truly national water grid, currently managing interbasin transfers of up to 500,000 m³/year. More than two thirds of water is consumed by Spain's heavily irrigated agricultural sector and Spanish households are relatively wasteful users of water, consuming twice the European average (AEAS National Water Supply Survey 2000). Very little of the water network is privatised although there is extensive private sector involvement in providing various services to the otherwise strongly publically-orientated water sector.

In this context Spanish water managers are both proud of their country's bold investment in the National Water Grid and worried about the future. Too many Spanish rivers experience extended periods of low flow already and climate change models suggest that this could get worse. The WWF is deeply concerned about the potential "desertification" of southern European nations like Spain, Italy and Greece (Deustche Welle, 2006). Ironically it may be the case that such extensive investment in hydrological infrastructure actually makes Spain more vulnerable as water users have come to expect the potent combination of government and civil engineering to always be able to come up with the water needed. Over the last few decades alternative water systems, especially for irrigation, have fallen by the wayside. One example is the acequia system of Andalucia; a system which once ensured that small amounts of water served a great many users and provided mechanisms for allocation when resources were particularly scarce.

The experience in Italy in the 1990s was in many respects like that of the postcommunist countries of Eastern Europe: restructuring followed by pressures to engage with foreign water multinationals yielding relatively little for consumers other than significantly higher prices (Lobina, 2008; OECD, 2004). With the territorial reorganisation of water services prices have been bound to a "costs plus" framework preparatory to attempts to attract private investment, although to date only a few Italian cities (e.g. Arrezo) have created joint ventures with foreign firms. Interestingly one restructured municipal firm, Rome's Acea has moved quickly to secure water concession contracts overseas including in Albania, Armenia, Bolivia, Colombia, Honduras and Peru.

Climate Change and European Hydro-futures

In this chapter it has only been possible to present the barest outline of the likely impacts of climate change on the status of water as a "resource" and as "hazard". One of the intriguing findings has been that water's dual identities may in fact be moving closer together such as the dividing line between resource and hazard is becoming tenuous indeed. At the level of the EU one set of initiatives specifically designed to help member nations adapt has been the Water Framework Directive (discussed in greater detail in Chapter 3 of this volume) – though we are only in the early stages of its implementation and already problems are becoming manifest (cf. Page and Kaika, 2003). Partly the problem is that the implementation principles of "subsidiarity" and "proportionality" mean that different member states may interpret even relatively simple EC directives quite differently from one another. But of course it is also the case that EC environmental policy is always hedged around with great complexities, uncertainties and even deliberate vagueness. Indeed there appear to be separate "camps" emerging reflecting different regional or national policy cultures – an issue that is currently the subject of further research by the author.

Suggested Activities for Further Learning:

1. Consider the projected implications of climate change for a European country or region. Remember that some regions may continue to receive roughly the same amount of precipitation, but falling according to a very different spatio-temporal pattern.

2. Consider the implications of climate change for "water as hazard", reflecting in particular on how greater resilience can be built into our "hydrosocial" systems.

3. Prepare a case study of a recent significant flood event, such as the July 2007 floods in England, the Boscastle flood of 2004 or the 1998 floods in Poland.

4. How can small islands like Malta or Cyprus adapt to even more diminished water resources?

Chapter 9

Managing Europe's Water: 21st Century Challenges

How can one round off a volume on such a broad topic as "the challenges to European water management in the 21st century"? I cannot help but feel that over the preceding eight chapters we have only scratched the surface. Though I have taken great pains to ensure that the information presented in those chapters is as relevant and up to date as possible I am haunted by all the topics that have *not* been treated so far. For example what about the issue of "embodied water" – the water contained in things and consumed by their production? Recent work by Dhayatker (2009) explores another dimension of the global water imbalance: the water that Western European countries are importing from water-stressed regions in the form of agricultural produce. Embodied water comes in at least two forms:

- The water contained in the thing itself – directly embodied water.
- The water it takes to produce something – indirectly embodied water.

Foodstuffs in particular have an obvious water content which is particularly important because, from the point of view of the producer, that water is used consumptively (that is to say it is not available for reuse). A single 100 gram orange, for example, may contain 50-60 millilitres of water directly and have required up to 70 litres of water to produce (Dhayatker, 2009; Hoekstra and Chapagain, 2006). Thus the roughly 150,000 tonnes of oranges imported into the UK in 2007 also involved the importation of 2.5 million litres of water. Since oranges are mostly grown only in regions already suffering from water stress the question arises: is this perhaps an inequitable transaction? As this simple example shows, the total volumes of directly embodied water being exported from water stressed areas is significant and demands attention.

Similarly there is the problem posed by bottled water. In the UK alone the market for bottled water has grown from about half a billion litres in 1993 to more than 2.2 billion litres in 2007, a growth rate of something like 4% per annum (Rothenbaugh, 2008). Since some of this water comes from water stressed regions again questions of equity can be raised – how fair is it to export scarce water from developing nations like Fiji when there is no sense in which the product can be said to be scarce in its destination market? Moreover, can the environmental costs of the booming trade in bottled water be justified when, as was pointed out in the previous chapter with respect to the proposed national water grid, long distance transport is neither economically nor environmentally justified? And does any of this make any sense when repeated blind taste testing shows that consumers do not reliably prefer bottled water over tap (Fox, 2009)?

I am aware also that so much more could have been said about individual European countries, yet there was simply not the space to give each country its own allotted pitch and, in any case, early on I elected to take a thematic and regional approach to organising this volume. It is thus very much a volume about European water management challenges even if much about specific European countries has not been said – it would have been interesting for example to consider further the special challenges facing island nations like Malta and Cyprus as well as archipelagos such as the Balearics, Las Canarias and the islands of the Aegean Sea.

Instead throughout this volume I have tried to emphasise the major trends in water management as they affect Europe and to point to some of the key challenges inherent in attempts to manage water more sensibly, more equitably and with due regard to the realities of climate change. Surely one of the "big" messages of this book is that European nations, long accustomed to thinking about water as a ubiquitous resource, are going to have to come to terms with the fact that relative scarcity will increase in the 21st century. Not only is the absolute availability of water likely to change, but we are constantly finding new uses for more water resources which implies that relative water availability may decline even more quickly. Of course even within Europe declining relative availability will be experienced differently in different nations and regions. The Mediterranean Basin will get drier and hotter even as parts of it, such as the Spanish Costas continue to grow in population. Central Europe will have to learn how to deal with declining snowpacks as well as altered precipitation patterns. Nations that rely on snowmelt to generate hydroelectric power, such as Norway and Iceland may have to revise their national energy strategies. And of course all European nations are going to have to think collectively about how to make a changing resource base "go further". The EU's Water Framework Directive is one bold step in the direction of collective resource management – though of course not all European countries are members of the EU.

Worryingly however, notwithstanding this collectivist thinking about water management there is still a pronounced tendency on the part of water managers to insist that water, or water services, should be privatised to the fullest extent possible. In Chapter 1 I presented the opposed ideologies of private resource management versus common property management and the theme of public versus private water management has woven its way through most of the chapters of this book. In Chapter 6 I explored the privatisation of water services across Europe much more directly, and found that the process raises as many problems as it solves. Indeed, there is now a movement towards "remunicipalisation" of water services in some parts of the world (Barlow, 2007). Paradoxically, whilst there remains strong political backing behind the water privatisation agenda – especially in the postcommunist nations of Eastern Europe – many of the successes of water management have been founded on the idea of water as a "commons". Participatory management of water at the scale of the catchment basin (mandated in, for example, the EC Water Framework Directive – discussed in Chapter 3) offers much potential for developing creative and socially-inclusive solutions to the water challenges of our time. Yet we will need to learn more about how the WFD and similar national-scale initiatives are actually implemented before we can judge their likely future success.

There is a profound paradox inherent in all that I have been saying here and that is that whilst it is collective action that creating the problems we have started to unpack in this volume, increasingly our response to these is framed in terms of individual responsibilities. So, an oft-proffered solution to declining relative availability of freshwater is to privatise water supply and sewerage and to impose water meters on consumers and businesses – this in spite of the clear fact that privatisation in Europe has been a very mixed blessing to say the least, and that there is very little research to support the contention that water metering either encourages conservation or augments socio-hydrological equity. In various parts of this book I have written at length about privatisation in Europe and about technical issues like water metering, but the point really is that in our neoliberal world the solutions to problems generated collectively are perceived to lie with individuals seen as rational economic actors. It is a long time since economic geographer Trevor Barnes deconstructed the myth of "homo economicus", one of whose key characteristics is that such conceptions

> involve turning away from the world. They reject the richness and diversity of human life by reducing it to some fundamental essence. In contrast to understand why people do the things they do, one must see human action within the broader context within which it occurs (Barnes, 1996, p. 98).

This may be somewhat abstract, so, to put it another way: "Privatisation formalised an important transformation in the underlying conceptions in the welfare state... through reclassifying their products as commodities. Our collective commitment to social justice has been fundamentally altered as a result." (Bakker, 2001, p. 157). There may be many different ways of delivering water services, arrayed along a continuum from completely private to wholly public, but surely any given strategy needs to be underpinned by a basic commitment to providing sufficient water each and ever citizen of our burgeoning world. Whilst it is still arguable whether or not the right to water proclaimed in the 1977 Mar del Plata Agreement and subsequent international documents is enforceable (cf. McCaffery, 1992) surely it is crystal clear that not providing this water is utterly unacceptable. How did we end up in the indefensible position of depriving much of the world of clean water through a set of arguments rooted in anti-state and pro-accumulationism?

It should be clear that what we are talking about here is the peculiarly postmodern, postfordist perversion of the modernist idea of human rights, in this case to water, into relative uses and needs which need to be properly calibrated in money cost terms so that competing uses can be optimised. Even purely environmental uses pose no problem, as long that they are corrected translated into cash equivalents and paid for. While it may seem sometimes that the anticommunist revolutions made little difference to the "engineering paradigm" prevailing in discourses about natural resource management, the renegotiation of the ideas of risk, cost and profit have subtly but profoundly transformed the situation. The reconfiguration of water resources as objects of privatised exploitation with lower levels of public sector intervention is expressive of nothing less than an epochal shift in environmental governmentality, defined by Foucault in terms of the implantation of specific naturalised ideals of "population," "political right," and "security" in the modern state:

> the setting in place of mechanisms of security...mechanisms or modes of state intervention whose function is to assure the security of those natural phenomena, economic processes and the intrinsic processes of population: this is what becomes the basic objective of governmental rationality (Foucault, 1991, p. 19).

Perhaps security is the inverse of risk? But I should like to emphasise that I am not at all saying that the physical nature of water resources, their spatial and temporal variability, issues of quality, etc. must not be taken into account, only that this "taking into account" still leaves us with much wider bounds of interpretation and understanding than we might initially think. As Latour points out:

> Knowledge does not reflect a real external world that it resembles via mimesis, but rather a real internal world, the coherence and continuity of which it helps to ensure (Latour, 1999)

What might all this mean for the central task of facing up to the challenges to European water management outlined in this book? Discourses about "sustainability" and "climate change" have conjoined in recent years to create a new popular consensus that European societies face a serious sustainability challenge in the first half of the 21st century. Put another way, whereas ten years ago sustainability was conceived largely in terms of a "steady state" ecological economics (e.g. Herbert Daly), growing realisation about the scale and speed of current climate change (whatever one thinks about causes) has changed the rules of the sustainability "game". The concept of sustainability currently remains as elusive as the search for a universal definition. However, it is widely agreed that sustainable development is an evolving process. "There are diverse approaches to sustainable development; some are very market-led and involve pricing nature while others involve putting environmental protection at the heart of policy and reducing consumption" (Willis, 2005 p. 201). From the ecocentric view, so-called deep ecologists argue for the rejection of even the "sustainable" utilisation of nature's assets – especially living resources (such as plants and animals). Technocentric opponents believe in the prospect of perpetual improvement in human standards of living through technological innovation. This argument was also articulated by the economist Julian Simon as he states that "the human brain is the ultimate resource which would solve all material obstacles to continued existence: necessity being the mother of all inventions" (Simon, 1981, cited in Murphy and Bendell, 1997

p. 62). These analytical positions can also be placed on a weak-strong continuum. "Strong sustainability places a premium on the peace of mind in knowing that one is living in truly sustainable world and involves full precautionary protection of ecosystems and maintenance of all natural capital" (O'Riordan, 1995 p. 22, cited in Myerson and Rydin, p. 103). Very weak sustainability "looks only to the overall stock of capital, natural and man made, and allows a degree of substitution between categories to maintain the overall stock level" (Myerson and Rydin, 1996, p. 104). Sadly, most European leaders (and citizens I daresay) clearly fall within the "weak" side of the spectrum.

Now, rather than a "steady state" conceptualisation of sustainability we are forced to consider how we maintain current standards of living with both rising populations *and* declining environmental stocks. Clearly collective action is called for, although as we have seen the dominant public policy paradigm that is comprised of "sustainability" on the one hand and "climate change" on the other seems to militate for a different sort of politics:

> In this era of generalised scarcity, everything has its price, and the degraded environment itself is the new commodity to be carved up, managed bought and sold. If an environmental amenity such a wetland or neighbourhood park is under threat from, say, a shopping mall or motorway, then prima facie there might be a case for compensation to be paid. But how should the value of the "loss" be assessed, and who pays and who should be paid? (O'Connor, 1993, p. 33).

These are the challenges – social, environmental, cultural, economic and political – that will define 21st century water management in Europe and in fact everywhere in our shrinking world.

Appendix 1
Glossary of Technical Terms

Term	Definition
µg/l	Microgram per litre.
Abstraction	The process of removing water from a supply system, whether it be natural (ground +surface waters) or engineered (reservoir, etc.) Abstraction can be consumptive or non.
Acid Mine Drainage (AMD)	Water draining out of surface or subsurface mines that has an extremely low pH (i.e. high concentration of hydrogen ions) and often contains high concentrations of sulfur, aluminum, and heavy metals such as iron.
Acre-foot	An antique English measure of volume sometimes still used in the US. 1 af = 1.182 m³.
Aerobic	Life processes occurring in the presence of free oxygen.
Alkalinity	A measure of the amount of anions of weak acid in water and of the cations balanced against them. See pH.
Anaerobic	Life processes that occur in the absence of free oxygen. See AEROBIC and ANOXIC.
Anthropogenic	Human caused or initiated changes to a natural system.
AONB	"Area of Outstanding Natural Beauty", a landscape conservation designation used in England and Wales.
Aquifer	Is a porous rock structure within which water travels and is stored. Aquifers may be shallow, a few metres in depth, or very deep being several hundred metres in depth.
Artesian	Water that arises to the earth's surface under hydrostatic pressure from an aquifer that is confined between two impermeable strata.
Artificial Water Body	Is a body of surface water created by human activity. It is known as a heavily modified water body if, as a result of physical alterations by human activity, it is changed substantially in character. Dam reservoirs are the most obvious, but not the only, example.
BAP	Biodiversity Action Plan.
Basin	An area in which the margins dip towards a common centre or depression, and towards which surface and subsurface channels drain. Cf. DRAINAGE AREA.
Aenthic	Of or pertaining to the bottom, or "bed," of a body of water.
Biochemical (Biological) Oxygen Demand (BOD)	The dissolved oxygen required to decompose organic matter in water and thus a measure of oxygen consumption in a fixed period of time. The standard methodology is to measure oxygen demand over 5 days at 20° C and is terms BOD_5. Unpolluted water usually measures less than 2 mg/l, raw sewage more than 500 mg/l.

Term	Definition
Biodiversity	An index of richness in a community, ecosystem, landscape and the relative abundance of these species.
Biofilm	Naturally occurring growth of micro-organisms on surfaces inside water mains.
Biota	living things such as plants, animals, and microorganisms.
Bog	Wet (poorly drained), spongy land with surface vegetation of mosses and shrubs, usually highly acid and nutrient-poor, with thick accumulations of poorly to finely decomposed peat.
Buffering Capacity	The ability of a solution to resist or dampen changes in pH upon the addition of acids or bases.
BW	British Waterways.
CAP	The Common Agricultural Policy of the European Commission.
Catchment	Something for catching water, as in a reservoir or basin; the water caught in such a basin. area the area of land drained within a watershed.
CCW	Consumer Council for Water, a statutory regulator of the privatised water industry in England and Wales.
Channel	A natural or artificial waterway of perceptible extent that periodically or continuously contains moving water, having a definite bed and banks which serve to confine the water.
Channelisation	The mechanical alteration of a stream which may include straightening or dredging of an existing channel, or creating a new channel to which the stream is diverted.
Closed Basin	1) A basin draining to some depression or pond within its area, from which water is lost only by evaporation or percolation. 2) A basin without a surface outlet for precipitation falling therein.
CMP	Catchment Management Plan.
Coastal Water	Is that area of surface water on the landward side of a line, every point of which is at a distance of one nautical mile on the seaward side from the nearest point of the baseline from which the breadth of territorial waters is measured, extending where appropriate up to the outer limit of transitional waters.
Coliforms	A group of generally harmless bacteria which may be faecal or environmental in origin.
Common Pool Resource	A natural resource that is the property of no-one, but which is usually of such magnitude that no potential user can be excluded (as opposed to "commons" which are usually collectively "owned" and managed). Air and surface waters are generally understood as common pool resources.
Condensation	The process by which a gas is cooled and converted into a liquid.
Consumption	Use of the resource that is "permanent" insofar as none is recycled. Water use is commonly divided into that which is consumptive and that which is non-consumptive.
Cryptosporidium	A water-borne protozoan parasite that increasingly commonly causes acute diarrheal illness in humans. It is occasionally found in UK water supplies as in the case of Mugdock reservoir near Glasgow in August 2002.

Term	Definition
CSO	Combined Sewer Overflow happens usually in periods of excess precipitation when the usually separate surface water and domestic/industrial effluent drainage systems are forced to discharge excess into the environment untreated.
Degradation	A decrease in water quality or, less commonly, quantity.
Demand Management Policies	Usually involving price mechanisms, that are designed to reduce consumer demand for water. Until the 1980s and 1990s much more common in water management was a focus on expanding supply – see the discussion of the Engineering Paradigm in Chapter 5.
Desiccation	Dehydrating or drying up.
Detritus	Fresh to partly decomposed plant or animal matter, often swept into surface water drainage or effluent systems and generally physically screened out.
Diffuse Sources (of pollution)	Primarily associated with run-off discharges related to different land uses such as agriculture and forestry, from septic tanks associated with rural dwellings, from landfill activities and from the application of fertilisers and pesticides in agriculture or silvaculture.
Dilution	Reduction in the concentration, or strength, of a substance in larger volumes of water. This is one of the ways that pollutants are dissipated or degraded in watercourses.
Discharge	Volume of water flowing in a given stream at a given place and within a given period of time, usually expressed as cubic meters per second or cubic feet per second.
Dissolved Oxygen (DO)	The concentration of oxygen dissolved in water, expressed in mg/l or as percent saturation, where saturation is the maximum amount of oxygen that can theoretically be dissolved in water at a given altitude and temperature. This is an absolute measure of oxygen availability as opposed to BOD, which is measure of demand for oxygen by organic matter in the water sample. These two quantities are commonly related via the well-known STREETER-PHELPS EQUATION (see also Chapter 4 and Appendix 3).
Dissolved Solids	Total of disintegrated organic and inorganic material contained in water.
Distribution Systems	A water company's network of mains, pipes, pumping stations and service reservoirs through which treated water is conveyed to consumers.
Drainage basin	See WATERSHED.
Drawdown	The lowering of water levels stored behind a dam or other control structure. Change in reservoir elevation during a specified time interval. Often drawdown patterns are cyclical and are related to diurnal or seasonal variations in incident precipitation and demand.
Drought	A period during which there is acute water shortage, usually caused by precipitation below the long-term average.
Dual Supply	The supply of water of different quality to different types of user (e.g. households versus industry).

Term	Definition
EC	European Commission, the executive body of the European Union.
EEC	European Economic Community.
Effluent	Anything that flows out of a source, such as sewage from a pipe.
El Niño	A periodic short-term climate change (range of 2-7 year intervals) due to warmer currents (as high as 14°C warmer) in the southern Pacific Ocean creating significant changes or reversals in barometric pressures in the Pacific and Indian Oceans (the "Southern Oscillation"). Together these warmer currents and changed barometric pressures are referred to as ENSO. The effects of ENSO are global, producing heavy rainfall or snowfall in some areas simultaneous with drought in other areas and therefore significantly affecting plant, animal and human well-being.
Embankment	A bank, mound, dike or like structure erected along or across a wetted area designed to hold back water or to carry a road/rail route.
Emission Controls	Are controls which are placed on emissions (point sources) requiring a specified emission limit value. See Emission Limit Value.
Emission Limit Value	Means the amount of a substance, usually expressed as a concentration and/or a level, which may not be exceeded during any one or more periods of time.
EN	English Nature.
Environmental Quality Standard	Means the concentration of a particular pollutant or group of pollutants in water, sediment or in aquatic life which should not be exceeded in order to protect human health and the environment.
Environmental Toxicology	The study of the movement of toxic compounds through ecosystems and their impacts.
Epilimnion	The upper water layer in a stratified lake that is more or less uniformly warm in the summer. It is separated from the lower, colder, water layer by the Thermocline.
EPT	An index of water quality based on the abundance of three pollution-sensitive orders of macroinvertebrates relative to the abundance of a hardy species of macroinvertebrate. It is calculated as the sum of the number of Ephemeroptera (Mayflies), Plecoptera (Stoneflies), and Trichoptera (Caddis Flies) divided by the total number of midges (Diptera: Chironomid).
ESA	Environmentally Sensitive Area – a scheme of the UK government to promote more environmentally sensitive agriculture in especially sensitive areas such as the Fenlands of Cambridgeshire and Lincolnshire and the Cotswolds of Gloucestershire. Replaced in 2005 by the Environmental Stewardship Scheme.
Estuary	1) The portion of a coastal stream influenced by the tide of the body of water into which it flows. 2) A bay, at the mouth of a river, where the tide meets the river current. 3) An area where fresh and marine waters mix.

Term	Definition
EU	European Union.
Euphotic Zone	The lighted region of a body of water that extends vertically from the surface to the depth at which light is insufficient to maintain light-requiring biological processes.
Eutrophication	The process of enrichment with nutrients, especially nitrogen and phosphorus, leading to increased production of organic matter. Eutrophication can also occur naturally as water bodies, especially lakes mature.
Evaporation	Is the transformation of liquid water into vapour as a result of heating. Its opposite is Condensation.
Evaporation Ponds	Ponds into which water is discharged and then allowed to evaporate in order to recover materials suspended or dissolved in the water. Historically evaporation ponds have been used as a way of "mining" sea salt, but can be used to remove other compounds or impurities from source water.
Faecal Coliforms	A sub-group of coliforms, almost exclusively, but not always, faecal in origin. Whatever their source they are a major threat to human health and therefore a primary target of water treatment processes.
Flood	Any water flow that exceeds the bank capacity (greater than bank discharge) of a stream or channel and flows out onto the flood plain. See FLOW. It is important to note that whilst flooding is generally seen negatively by humans, it is an unavoidable and indeed beneficial natural phenomenon.
Flood Plain	Area adjoining a body of water that may become inundated during periods of maximum water levels. In the 19th and 20th centuries many flood plains were built over or otherwise re-engineered such that natural absorptive capacities of river courses were compromised.
Flow	The movement of water, and the moving water itself. The volume of water passing a given point per unit of time. See Discharge. Note: "Base Flow" is the portion of stream discharge that is derived from natural storage. Subsurface Flow is that portion of the groundwater moving horizontally through and below the stream bed. Surface Flow is that part of water in a channel flowing on the surface of the substrate.
Fluvial	Pertaining to streams or produced by stream action, from the Latin "fluvius" meaning river.
Fossil water	Groundwater trapped under impermeable rock which can be thousands of years old and tends to be heavily mineralised.
Giardia	A waterborne protozoan parasite causing illness in humans.
Groundwater	Is all water which is below the surface of the ground in the saturation zone and in direct contact with the ground or subsoil. This zone may be referred to as an "aquifer" if it is of sufficient porosity and permeability to allow a significant flow of groundwater or the abstraction of significant quantities of groundwater through well-building. See Fossil Water.
Groundwater Budget	The summation of the movement of water into and out of a groundwater reservoir during a specified period of time.

Term	Definition
Groundwater Discharge	Flow of water from the groundwater reservoir onto the earth's surface.
Halophyte	A plant that is more or less restricted to saline soil or to sites that are influenced by salt water, some of which are now used in environmental remediation work.
Headwaters	The upper tributaries of a drainage or catchment basin.
Hydraulic Gradient	(1) The slope of the water flowing in a watercourse. (2) The drop in HYDRAULIC HEAD per length in the direction of stream flow.
Hydraulic Head	The height of the free surface of a body of water above a given point beneath the surface, or the height of the water level at an upstream point compared to the height of the water surface at a given location downstream.
Hydraulic Society	An agrarian society based on large scale hydraulic constructions. Understood by many, following the German sociologist Karl Wittfogel (1896-1988), to have precipitated the development of a centralised state and urban society.
Hydrological	Means pertaining to water.
Hydrological Cycle (sometimes called the Water Cycle)	Is the movement of water from the atmosphere to the land as precipitation, its flow to the sea via rivers and lakes and the return to the atmosphere via evaporation.
Hydrophyte	A plant which has evolved with adaptations to live in aquatic or very wet habitats, e.g. cattail, water lily, water tupelo.
Hydrosocial Cycle	The socio-natural process of freshwater abstraction, storage, treatment, distribution, use; and wastewater collection, treatment and disposal.
Hydrosphere	The area in which water exists; for the purpose of this module, this sphere includes all liquid water on Earth, such as rivers, lakes, and oceans, as well as all frozen water, such as glaciers, icebergs, and polar icecaps.
Hypolimnion	The bottom layer of a stratified water body that is isolated from the atmosphere and more or less cold in the summer – that portion lying below the Thermocline.
Indicator Organism	An organism whose presence or absence indicates the presence of contamination and hence the possible presence of pathogens.
Infiltration	The process by which water moves from a surface body into the groundwater system.
Inland Water	Is all standing or flowing water on the surface of the land (such reservoirs, lakes, rivers and coastal waters) and all groundwater on the landward side of the baseline from which the breadth of territorial waters is measured.
Karst Springs	Water forced up from subterranean reservoirs through "cracks" in the overlaying limestones (which are often created through solution of the rock by contact with water). This water tends to be high in "temporary hardness", that is to say unresolved calcium carbonates $CaCo_3$.
Lagoon	An area of shallow water separated from the sea or other larger water body by a low bank. Also, a settling pond for treatment of wastewater.

Term	Definition
Lake	Is a body of water, which may be man-made or natural, occurring on the land surface.
Littoral	The onshore area of a water body, extending from the shore to the limits of rooted aquatic plants. See also Pelagic.
LTRQOs	Long Term River Quality Objectives, a tool used by the Environmental Agency for England and Wales.
m³/d	Cubic metres per day, a measure of water supply.
m³/s	Cubic metres per second.
Macroinvertebrates	Small, spineless creatures that are visible with the unaided eye; they include organisms such as crustaceans, mollusks, worms, and insects. Often used as proxy measures of water quality. See ETP.
Marsh	A water-saturated, poorly drained wetland area, periodically or permanently inundated to a depth of up to 2 m (6.6 feet) that supports an extensive cover of emergent, non-woody vegetation, essentially without peat-like accumulations (which would make it a FEN).
Metalimnion	The water layer of a stratified lake between the epilimnion and the hypolimnion characterised by a steep thermal gradient.
mg/l	Milligram per litre.
Mitigation	The technically driven attempt to repair damage to a human or natural (or SOCIONATURAL) system without necessarily solving the problem which caused the damage in the first place.
Ml	Megalitre.
Ml/d	Megalitres per day.
Navigable	Generally, waters used by humans for transportation. Legally, interstate waters; intrastate lakes, rivers and streams which are utilised by interstate travellers for recreational or other purposes; intrastate lakes, rivers and streams from which fish or shellfish are taken and sold in interstate commerce; and intrastate lakes, rivers and streams which are utilised for industrial purposes by industries in interstate commerce.
Non-consumptive Use	That portion of water use that is not "permanently" used, i.e. it is returned to the hydrosocial cycle. See Consumption.
Non-point-source Pollutant	A pollutant that cannot be traced back to one specific source, but instead comes from a general area, such as runoff from a parking lot or agricultural field.
Nutrient	An inorganic (i.e. non-carbon-based) element required by an organism for normal growth and activity.
OFWAT	Office of Water Services, the primary economic regulator of the water industry in England and Wales.
Outwash Plain	A flat area of land formed by sediment carried to the site from a glacier and deposited by changes in stream carrying capacity.
PAH	A group of about a dozen characteristic organic compounds known as polycyclic aromatic hydrocarbons and primarily formed by incomplete combustion. Many of these are known or suspected carcinogens.

Term	Definition
Parameter	The measure or indication of something, such as "water quality parameters".
Parts Per Million (ppm)	Unit of measure most often used to describe the amount of a particular gas or compound in the air or water. It is the proportion of the number of molecules of the gas or compound out of a million (1,000,000) molecules of air or water.
Pathogen	Anything that causes disease.
Pelagic	Open water areas away from the shore and the bottom. Cf. Littoral.
pH	A logarithmic scale ranging from 1 to 14 that identifies the acidity of a solution as the negative logarithm of the concentration of hydrogen ions in solution. Because the pH scale is logarithmic, each unit change represents a ten-fold change in the concentration of hydrogen ions. (pH=7.0 is neutral).
Photosynthesis	The process by which green plants convert solar energy into chemical energy in the form of organic (carbon-containing) molecules, releasing oxygen as a by-product; $6\ CO_2 + 6\ H_2O +$ sunlight $C_6H_{12}O_6 + 6\ O_2$.
Plankton	Microorganisms suspended in the water having little or no power of locomotion, carried by waves, currents and other movements of water. Phytoplankton: microscopic floating aquatic plants; Zooplankton: microscopic floating aquatic animals.
Point-source Pollutant	A pollutant that can be traced back to one specific source, such as sewage from a pipe.
Polluter Pays Principle	Requires that the polluter of the water environment should wherever possible pay for the remediation or mitigation of their pollution.
Potable	Suitable or safe for drinking.
PPG/PPS	Planning Policy Guidance/Planning Policy Statement notes are outline principles for guiding/constraining the planning process in England and Wales. For our purposes PPS 23, on water and air quality, and PPG 25 on flooding are the most relevant.
Precipitation	The process by which liquid or solid water (rain, sleet, snow, etc.) moves from the atmosphere to Earth's surface.
Qanat	A water supply system that taps groundwater and directs it through gently sloped tunnels and channels over a wide area. The longest qanat gallery is in central Iran and is over 80km.
Rain Shadow	An area to the leeward of a high land mass, particularly a mountain range, which receives less rain than would be expected had the high land mass not been upwind of it. NOTE: What it receives is mainly the residue of orographic rain.
Remediation	The process by which something is corrected, in contrast to merely mitigated See Mitigation.
Reverse Osmosis Filtration	A system that depends on osmotic pressure to remove salt from seawater, thus making it drinkable. It is relatively high maintenance, slow and expensive.
RHS	River Habitat Survey.

Term	Definition
Riffle	Portion of a stream characterised by fast-moving, turbulent, relatively shallow water flowing over a substrate of rocks of various sizes.
Riparian	Terrestrial areas adjacent to perennial or intermittent water bodies that provide functional linkages between terrestrial and aquatic ecosystems through coarse and fine organic matter input, bank stability, water temperature regulation, sediment and nutrient flow regulation, and maintenance of unique wildlife habitat.
River	Is a body of inland water flowing for the most part on the surface of the land but which may flow underground for part of its course. Upland rivers are generally fast flowing and lowland rivers are generally slow flowing and meandering.
River Basin	Means the area of land from which all surface water run-off flows, through a sequence of streams, rivers and lakes into the sea at a single river mouth, estuary or delta.
River Basin District	Is a river catchment or a group of catchments defined as a planning unit for the purposes of the EC Water Framework Directive.
River Basin Management Plan	Is a detailed document describing the characteristics of the basin, the environmental objectives that need to be achieved and the pollution control measures required to achieve these objectives through a specified work programme.
Runoff	Precipitation that has drained through or over an area, often taking on loads of pollutants or sediments.
SAC	Special Area of Conservation are areas of strict environmental protection and are defined and gazetted under the EC Habitats Directive.
Sediment	Small pieces of organic or inorganic material that are deposited on Earth's surface by wind, water, or ice.
Sewage	waste water from residential, industrial or other activities.
Siltation	The movement of silt – tiny particles of clay and sand – into streams during erosion.
SSSI	Site of Special Scientific Interest is a conservation designation mandated in the 1981 Wildlife and Countryside Act of England and Wales.
STW	Sewage Treatment Works.
Substrate	The materials, such as sand, gravel, or cobble, that make up the bottom, or "bed," of a body of water.
Succession	The gradual supplanting of one community of plants by another.
Surface Water	Means inland waters, except groundwater, which are on the land surface (such as reservoirs, lakes, rivers, transitional waters, coastal waters and, under some circumstances, territorial waters) which occur within a river basin.
Suspended Solids	Small pieces of organic or inorganic material floating in a column of water.
SWQO	Statutory Water Quality Objectives were mandated in the Water Resources Act 1991 for England and Wales (updated in 1999).

Term	Definition
Thermocline	The zone of water depth within which temperature rapidly changes between the upper warm water layer (Epilimnion) and lower cold water layer (Hypolimnion).
THM	A group of organic substances known as trihalomethanes which are both a pollutant and a common byproduct of the water disinfection process.
Tributary	A stream that flows into, or "feeds," another stream.
UWWTD	Urban Wastewater Treatment Directive.
VOC	Volatile Organic Compounds are chemical compounds related to industry and which cause significant atmospheric and water pollution.
Water Body	Is a discrete and significant element of surface water such as a river, lake or reservoir, or a distinct volume of groundwater within an aquifer.
Water Budget	The balance of all water moving into and out of a specified area in a specified period of time.
Water Shortage	Is generally said to exist in places where total availability is less than 1,000 m^3/p/yr, or approximately 2,740 l/p/day for all purposes.
Water Stress	Is generally said to exist in places where total availability is between 1,000 and 2,000 m^3/p/yr
Water Table	Depth below which the ground is saturated with water.
Watershed	1) The region or area drained by a stream or river. 2) Can also include regions and small watersheds in which rivers/streams lead into larger bodies of water such as certain lakes. See also Catchment.
Weir	A barrier constructed across a stream to divert fish into a trap or to raise the water level or divert its flow, with water flowing over the top of the structure.
Wetland	Land where saturation with water is the dominant factor determining the nature of soil development and the types of plant and animal communities living in the soil and on its surface. Soil or substrate that are at least periodically saturated with or covered by water, and differ from adjoining non-inundated areas. A general term for any poorly-drained, uncultivated tract, whatever its vegetation cover and soil.
Zooplankton	Microscopic animals living within the water column Cf. Plankton.

Appendix 2
Water Quality Parameters

pH

The scale which is used to describe the concentration of acid or base is known as "pH", for power or potential of the Hydrogen ion. A pH of 7 is neutral. pHs above 7 are alkaline (basic); below 7 are acidic. The scale runs from about zero, which is very acidic, to fourteen, which is highly alkaline and is logarithmic, meaning that each change of one unit of pH represents a factor of 10 change in concentration of hydrogen ion. So a solution which has a pH of 3 contains 10 times as many (H+) ions as the same volume of a solution with a pH of 4, 100 times as many as one with a pH of 5, a thousand times as many as one of pH 6, and so on.

Some common materials and their approximate pH's are:

Acids: carbonated beverages, 2 to 4; lemon juice, about 2.3; vinegar, about 3;
Bases: baking soda, 8.4; milk of magnesia.10.5; ammonia,11.7;lye,14 to 15.
(Some of these figures are from the Handbook of Chemistry and Physics, 56th ed., CRC Press,1976.)

Within a pH range of between 5.5 and 8.5 is considered tolerable for most aquatic life.

Dissolved oxygen

Dissolved oxygen is one of the best indicators of general water quality. As a general rule, the higher the DO, the better the water quality. If possible, the dissolved oxygen test should be done on site. Dissolved oxygen is dependent on temperature: warm liquids hold less dissolved than cold liquids. Also, when organic wastes decompose in a body of water, dissolved oxygen is used up. Because more aquatic organisms are cold blooded, their metabolism rises as temperature goes up and the amount of available oxygen goes down. This often results in fish "kills", especially if the DO drops below 5 ppm.

Like solids and liquids, gases can dissolve in water. And, like solids and liquids, different gases vary greatly in their solubilities, i.e. how much can dissolve in water. A solution containing the maximum concentration that the water can hold is said to be saturated. Oxygen gas, the element which exists in the form of O_2 molecules, is not very water soluble. A saturated solution at room temperature and normal pressure contains only about 9 parts per million of D.O. by weight (9 mg/L). Lower temperatures or higher pressures increase the solubility, and vice versa.

Alkalinity

Alkalinity is a measure of all of the substances in water which have the ability to react with the acids in water and "buffer" the pH; i.e. the power to keep the pH from changing. Pure water would have a pH of 7 and therefore has no (zero) alkalinity. Alkalinity is important for aquatic life because it protects or "buffers" much as a Tums or Alka Seltzer does in your stomach: it keeps the pH from changing and makes the water less affected by factors such as acid rain or acid spills. The main sources of alkalinity in water are rocks which contain carbonates or bicarbonates and respiration.

Limestone is a good example of this type of rock. Water with a total alkalinity of 100-120 ppm is considered to be the best waters for fish and aquatic organisms. Lakes with a total alkalinity of below 50 ppm are considered "too clean" and will support little life while lakes of 200 ppm are quite high. Alkalinity can be increased in water by adding lime to it. Alkalinity is considered a fairly reliable measure of the productivity of a body of water.

Total alkalinity – less that 50 ppm – too low
200 ppm – becoming too high
100-120 ppm – best waters for fish with a pH between 7-8

Carbon dioxide

As green plants carry on photosynthesis only in the presence of sunlight, they as well as the animals in the water carry on respiration at night; therefore more carbon dioxide tends to build up in water during the night rather than in the daytime. If carbon dioxide levels are high and dissolved oxygen levels are low, fish have trouble carrying on respiration and their problems are worse if the water temperature rises. Fortunately for fish, "free" (uncombined) carbon dioxide rarely exceeds 20 ppm and most fish can survive this amount with no ill effects. Higher than this for any length of time can be lethal.

Surface waters:
1.0-6.0 ppm – fish avoid these waters
6.0-12.0 ppm – ideal for most fish
Above 12 ppm – few fish can survive for long periods of time

Hardness

Hardness in water is the amount of calcium and magnesium in the water. The natural source of hardness is usually limestone rock. The most frequent test performed on water is for total hardness. Hardness is important to living organisms because soft water makes heavy metals such as mercury and lead more poisonous to fish. Some nonmetals such as ammonia and certain acids are also more toxic to fish in soft water. There is some evidence that humans who drink soft water over a long period of time are more susceptible to cardiovascular disease. Vertebrates need calcium to build bones. All living things need calcium and magnesium in order for proper cell functions. Chlorophyll contains magnesium. Drinking water with a total hardness of 250 ppm is best. Over 500 ppm can make you very ill.

Soft: 0-50 ppm
Moderately hard: 60-120 ppm
Hard: 121-180
Very hard: over 180 ppm

Nitrates/nitrites

Nitrogen compounds are essential for healthy plant growth. Nitrogen is a major constituent of commercial fertiliser. The presence of excessive amounts of nitrogen compounds in water supplies presents a major pollution problem. Large amounts of nitrates and nitrites in water are harmful to humans. Nitrates in conjunction with phosphates can cause algal blooms. The US EPA states that 10ppm of nitrate/nitrogen is a limit that should not be exceeded. Lower amounts are desirable, but in some areas of the UK nitrate concentration in water is in the region of 100 ppm.

Nitrites, which are more toxic, are present primarily as the result of microbial reduction of nitrates. Associations between nitrates and cancer have been extensively investigated (McDonald and Kay, 1988).

Nitrates:
1. Trace amounts present in most natural waters – less than 1 ppm
2. Lakes with total nitrogen over 30 ppm susceptible to algal bloom
3. 10 ppm or less – acceptable for drinking water – higher is a health or environmental hazard

Nitrites:
1. Levels as low as 0.3 ppm can be harmful to fish
2. 1 ppm or above should not be used for feeding babies – could lead to "blue baby syndrome"

Phosphates

Primary sources of phosphates are agricultural run-off and detergent residue in urban wastewater. Phosphates in water along with the presence of phosphates stimulates the growth of algae. This in turn can lead to accelerated eutrophication of a body of water. The concentration of phosphates in water is normally not more than 0.1 ppm unless the water has been polluted.

Biochemical Oxygen Demand (BOD)

Biochemical oxygen demand, abbreviated as BOD, is a test for measuring the amount of biodegradable organic material present in a sample of water. The results are expressed in terms of the mg/L of DO which microorganisms, principally bacteria, will consume while degrading these materials.

For reasons discussed earlier, the depletion of oxygen in receiving waters has historically been regarded as one of the most important negative effects of water pollution. Preventing these substances from being discharged into our waterways is a key purpose of wastewater treatment. Monitoring BOD removal through a treatment plant is necessary to verify proper operation. However, because the test takes too long to be useful for short-term control of the plant, the chemical or instrumental surrogate tests are often used as guides. Unpolluted water =< 2 mg/l 02; raw sewage ~ 600 mg/l 02; treated sewage effluent ~ 20-100 mg/l 02.

Coliform bacteria

Another excellent indicator of water quality is based on the number of coliform bacteria. Coliform bacteria normally live in the intestines of mammals and are excreted with the faecal wastes. Some forms are pathogenic but even if they are not, if they are present in a water sample this indicates that the sample has been through an animal intestine.

Sewage contains large numbers of microbes which can cause illness in humans, including viruses, bacteria, fungi, protozoa and worms (and their eggs or ova). They originate from people who are either infected or are carriers. While many of these can be measured directly by microscopic techniques (some after concentration), the analyses most commonly performed are for so-called "indicator organisms." These organisms, while not too harmful themselves, are fairly easy to test for and are chosen because they indicate that more serious pathogens are likely to be present. For instance, wastewater treatment plants are often required to test their effluents for the group known as "faecal coliforms," which include the species E. coli, indicative of contamination by material from the intestines of warm-blooded animals. Water supplies test for a more inclusive group called "total coliforms", and in some cases, for general bacterial contamination (heterotrophic plate count, or HTP).

Besides, direct observation, identification of pathogenic microorganisms can be done by standard techniques used in clinical laboratories involving observing reactions in a battery of different indicating media. Some newer methods use chromatography to identify patterns of compounds which serve as "fingerprints" for certain bacteria; DNA analysis is another recent innovation. Most wastewater treatment plants, however, confine their testing to simply counting the numbers indicator bacteria.

Appendix 3
Water Quality Exercises

Exercise 1: Biodegradation of Pollution

Have a look at the following exercise, which illustrates the processes of biodegradation of pollution which take place in rivers. You will also learn a simple method of modeling surface water quality. Subsequently you may elect to use these water quality modeling techniques to determine the scope of activities/ interventions required to achieve a satisfactory level of river water quality.

Domestic wastewater is discharged into river AA at point R. Due to the nature of the pollution (easily degradable organic substances) we describe water quality with the use of the *biochemical oxygen demand* parameter (BOD_5) and the amount of dissolved oxygen (DO) in water. Consider the following parameters:

- Rate of river flow at the wastewater discharge point is $Q_R = 9.0$ m³/s,
- ambient BOD_5 at this point equals $C_R = 6.0$ mg/l,
- oxygen concentration at this point is $DO_R = 8$ mg/l,
- mean water velocity in the studied stretch of the river is $v = 0.2$ m/s,
- wastewater discharge outflow is $Q_S = 0.5$ m³/s and BOD_5 value is $C_S = 300$ mg/l.

Calculate the modification of the BOD_5 and DO values along the run of the river AA to a total length of $L = 50$ km downstream from the discharge, in steps of $L = 1$ km, on the following assumptions:

- wastewater is completely intermixed with river water immediately downstream of the discharge;
- pollutants undergo biological decomposition with the run of the river in aerobic conditions; we will disregard situations in which due to discharge of excessive loads of pollutants into the river there are periodical dissolved oxygen deficits, hence slowing down the decomposition processes;
- we will not take into account pollution sedimentation or adsorption;
- the process of biodegradation is described by the simplest decomposition equation, proposed by Streeter – Phelps (Dunn et al. 1992) where pollution concentration modifications depend on its initial concentration, duration of decomposition process and the reaction rate coefficient:

$$Ct = C0 * e -k1 * t$$

where:

C_t –   pollutant concentration after a period t [mg/l],
C_0 –   initial pollutant concentration [mg/l],
t –   duration of decomposition process [day],
k_1 –   constant decomposition rate [1/day].

- modifications of dissolved oxygen deficiencies (defined as the difference between the maximum dissolved oxygen concentration DOmax at the given temperature and the current concentration DO [mg/l]) are caused by uptake of oxygen by microorganisms during decomposition of pollutants and penetration of atmospheric oxygen into water (the rate of such penetration – reaeration – depends on the current oxygen deficit and the constant rate of reaeration); they are described by an equation:

$$DDOt = k1/(k2\text{-}k1) * C0 * (e^{-k1* t} - e^{-k2* t}) + DDO0 * e^{-k2* t}$$

where:

DDO_t –   oxygen deficit after time t [mg/l],
DDO_0 –   initial oxygen deficit [mg/l],
k_2 –   constant reaeration rate [1/day].

Accept the following data for computations:

DO_{max} =   11.0 [mg/l],
k_1 =   0.50 [1/day].
k_2 =   1.15 [1/day].

Present your computation results in graphic and tabular formats. Note that this is a moderately complex set of calculations, but it can be modeled as an equation using any standard spreadsheet package such as Microsoft Excel, which will also produce the requisite graphs.

Appendix 4

Analysis of the Water Supply Opportunities for an Industrial Plant

Construction of an industrial plant is planned immediately downstream of the water gauge section; the water demand of this plant is constant in time and amounts to 2.4 m³/s. (Note: that we are not making any claims about the precise nature of the demand at this stage. In reality, any application or abstraction would have to specify, at the very least, the amount to be abstracted, the amount used consumptively, the amount returned to the watercourse and the qualities of that water. Examine the potential for ensuring water supply for the plant from the river resources based on the following data:

1. *1.n*-year sequence of mean periodic flows $(Q_{ij} [\text{m}^3/\text{s}]; i=1,2,\ldots,n; j=1,2,\ldots,m)$ where:

    n – number of years;
    m – number of periods into which a year was divided; in the discussed example $n = 1$ year, $m = 12$ months;
    Data on mean monthly flows are presented in Table 5.1.
2. level of *base flow* QN [m³/s] along the river stretch downstream of the industrial plant water intake ($QN = 0.50$ m³/s);
3. *time guarantee* G_{kr} [-] required by the plant for meeting their water needs ($G_{kr} = 0.80$).

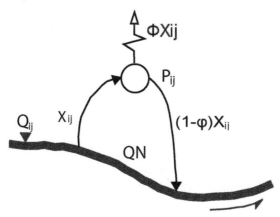

Figure A4.1 Diagram of water system component location

Designations and Definitions:

Length of time interval (t_{ij}) defines the number of million seconds in *j*-period of *i*-year.

Available water resources (QD_{ij}) define the volume of water, which may be drawn by the user to meet his/her needs in *j*-period of *i*-year:

$$QD_{ij} = \max\{0, Q_{ij} - QN\}$$

Water intake by the user (X_{ij}) in *j*-period of *i*-year:

$$X_{\downarrow}ij = \min\{P_{\downarrow}ij, [\![QD]\!]\downarrow ij\}$$

Water deficit suffered by the user (D_{ij}) in j-period of *i*-year:

$$D_{\downarrow}ij = \max\{0, [\![P_{\downarrow}ij - X]\!]_{\downarrow}ij\}$$

Time guarantee of meeting the user's water demand defines the relation of the number of time intervals during which the water supply task was implemented to the number of periods in which demand was declared,

$$G_v = 1 - \frac{\sum_{i=1}^{n}\sum_{j=1}^{m}\partial_{ij}}{\sum_{i=1}^{n}\sum_{j=1}^{m}Y_{ij}}$$

$$\partial_{ij} = \begin{cases} 1 & \text{when } D_{ij} > 0 \\ 0 & \text{when } D_{ij} \leq 0 \end{cases} \qquad \gamma_{ij} = \begin{cases} 1 & \text{when } P_{ij} > 0 \\ 0 & \text{when } P_{ij} = 0 \end{cases}$$

where:

$\sum_{i=1}^{n}\sum_{j=1}^{m}\partial_{ij}$ — number of periods in which the water supply task was not fully implemented,

$\sum_{i=1}^{n}\sum_{j=1}^{m}Y_{ij}$ — number of periods in which water demand was declared.

Volumetric guarantee of meeting the user's water demand defines the ratio of total volume of water supplied to the user to total volume of declared demand in all periods of the studied multi-annual period.

$$G_v = \frac{\sum_{i=1}^{n}\sum_{j=1}^{m}X_{ij}}{\sum_{i=1}^{n}\sum_{j=1}^{m}P_{ij}} \times \Delta t_{ij}$$

Table A4.1   Analysis of the possibility of water supply for the industrial plant from river resources

Month	Δ tij [mill. s]	Qij [m³/s]	QDij [m³/s]	Qdij Δ tij [mill. m³]	Pij [m³/s]	Pij Δ tij [mill. m³]	Xij [m³/s]	Xij Δ tij [mill. m³]	Dij [m³/s]	Dij Δ tij [mill. m³]	Δ ij
1	2.678	2.70	2.20	5.89	2.4	6.43	2.20	5.89	0.20	0.54	1
2	2.419	3.20	2.70	6.53	2.4	5.81	2.40	5.81	0,00	0.00	0
3	2.678	3.30	2.80	7.50	2.4	6.43	2.40	6.43	0,00	0.00	0
4	2.592	2.80	2.30	5.96	2.4	6.22	2.30	5.96	0.10	0.26	1
5	2.678	2.40	1.90	5.09	2.4	6.43	1.90	5.09	0.50	1.34	1
6	2.592	3.10	2.60	6.74	2.4	6.22	2.40	6.22	0,00	0.00	0
7	2.678	3.50	3.00	8.03	2.4	6.43	2.40	6.43	0,00	0.00	0
8	2.678	2.50	2.00	5.36	2.4	6.43	2.00	5.36	0.40	1.07	1
9	2.592	2.80	2.30	5.96	2.4	6.22	2.30	5.96	0.10	0.26	1
10	2.678	3.60	3.10	8.30	2.4	6.43	2.40	6.43	0,00	0.00	0
11	2.592	2.70	2.20	5.70	2.4	6.22	2.20	5.70	0.20	0.52	1
12	2.678	2.40	1.90	5.09	2.4	6.43	1.90	5.09	0.50	1.34	1
Sum	31.500	35.00	29.00	76.15	28.8	75.68	26.80	70.36	2.00	5.33	7
Average	2.630	2.92	2.42	–	2.4	–	2.23	–	0.17	–	–

Notes: Time guarantee Gt = 0.42; Volumetric guarantee Gv = 0.93.

In accordance with the above definitions and formulas Table A4.1 presents computations for all the studied time intervals (12 months): the volume of available water resources, the volume of water supplied to the industrial plant and the level of water deficit in the plant. The value of time guarantee of meeting the user's water needs was also computed.

We may see in Table A4.1 and Fig. A4.2 that during the studied 12 months water deficit occurred 7 times in the plant, i.e. the time guarantee of meeting the user's water demand is Gt = 0.42. It is much lower than the guarantee required by the user (Gkr = 0.80).

A problem has appeared: the available water resources in the river are insufficient to satisfy the water demand declared by the user with the required guarantee $G_{kr} = 0.80$.

$$\sum_{i=1}^{n=1} \sum_{j=1}^{m=1_2} QD_{ij} \times \Delta t_{ij} = 76.15mln\ m^2$$ At the same time it should be noted that the sum of available water resources during the studied period exceeds the sum

$$\sum_{i=1}^{n=1} \sum_{j=1}^{m=1_2} P_{ij} \times \Delta t_{ij} = 75.68mln\ m^2$$ of water demand in that period.

This means that total available resources allow meeting the total water demand but their distribution in time is unfavorable and they are featured by high variability.

Question: What can be done in such a situation? Before you go on to the next part of the case study consider the problem for a while.

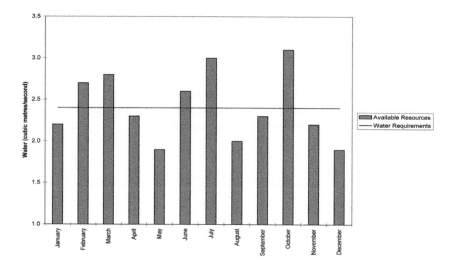

Figure A4.2 Comparative analysis of available resources and water demands

Possible solutions:

1. Increasing the available water resources by construction of a *storage reservoir* upstream of the industrial plant water intake;
2. Increasing the available water resources by construction of a *transfer channel* from another catchment, feeding the river upstream of the industrial plant water intake;
3. Increasing the available water resources by construction of an underground water intake;
4. Reducing the water demand of the industrial plant by changing the production technology.

The learned and rational choice of one of the above mentioned alternative solutions (or a variant which is a combination of solutions) requires completion of additional simulation computations focused on definition of parameters of hydrotechnical facilities in each alternative (*usable capacity* of storage reservoir, *capacity of transfer channel*, efficiency of underground water intake) and the required degree of water demand reduction.

The discussed example related to the simplest imaginable situation:

a. only one user in the system;
b. the degree of water supply is featured by a single criterion (time guarantee);
c. analyses were based on a historical flow series.

The next stage of this exercise would, most likely, involve conceptualisation of a storage reservoir and water channel system. Some of these issues will be picked up in the chapter on "Dams", but readers are warned that such considerations quickly become more properly the aegis of civil engineering.

Bibliography

Africa News, 2006 "South Africa: Water for All?", *Africa News*, 22 December 2006.

Africa News, 2007a "Nigeria: Water Supply – Government to Introduce Pre-Paid Meter", *Africa News*, 12 October 2007.

Africa News, 2007b "South Africa: Flood of Anger Over Johannesburg Water Plan", *Africa News*, 10 May 2007.

Agnew C, Anderson E, 1992 *Water Resources in the Arid Realm*, London: Routledge.

Ahmad Y.J, Serafy S, Lutz E, 1989 *Environmental Accounting for Sustainable Development,* Washington, World Bank, A UNEP – World Bank Symposium.

Alcamo J, Dronin N, Endejan M, Golubev G, Kirilenko A, 2007 "A new assessment of climate change impacts on food production shortfalls and water availability in Russia", *Global Environmental Change,* In Press, Corrected Proof.

Alcamo J, Henrichs T, 2002 "Critical Regions: A model-based estimation of world water resources sensitive to global changes", *Aquatic Sciences* 64(4), pp. 352-362.

Allan T, 2001 *The Middle East Water Question: Hydropolitics and the Global Economy*, IB Tauris and Co.

Allouche J, Finger M, 2001 "Two Ways of Reasoning, One Outcome: The World Bank's evolving philosophy in establishing a 'sustainable water resources management' policy", *Global Environmental Politics* 1(2), pp. 42-47.

Amos J, 2008 "Swiss glaciers 'in full retreat'", *BBC Online*, 15 December 2009.

Anderson S.C, Tabb B.H, (eds) 2002 *Water, Leisure and Culture: European historical perspectives*, Oxford: Berg.

Anderson T.L, 1983 *Water Crisis: ending the policy drought*, Johns Hopkins University Press.

Anderson T.L, Snyder P, 1997 *Water Markets: priming the invisible pump*, Cato Institute.

Aparcio J, Hidalgo J, 2004 "Water Resources Management at the Mexican Borders," *Water International* 29:362-74.

Appleby P, 2006 "Water Use", *Regeneration Magazine*, April.

Ashworth W, 1982 *Nor Any Drop to Drink*, Summit Books.

Atampugre N, 1997 "Aid, NGOs and Grassroots Development: Northern Burkina Faso," *Review of African Political Economy* 71 (1997): pp. 57-73.

Author Unknown, 2007 "Meter Consensus Out of Reach," *Kamloops Daily News*, 23 October, p. A6.

Bachelard G, 1942 *L'eau et les Rêves: essai sur l'imagination de la matière*, Paris: José Corti.

Bakker K, 1996 "Yorkshire Water and Yorkshire's Water: flows of water/capital in the drought of 1995", Working Paper #98/01, Economic Geography Research Group.

Bakker K, 2001 "Paying for Water: water pricing and equity in England and Wales", *Transactions, Institute of British Geographers* NS 26, 143-164.

Bakker K, 2005 "Neoliberalizing Nature? market environmentalism in water supply in England and Wales", *Ann.Assoc.Am.Geogr.,* 95(3), 542-565.

Balanyá B, Brennan B, Hoedeman O, Kishimoto S, Terhorst P, (eds) 2005 *Reclaiming Public Water: Achievements, Struggles and Visions from Around the World, Transnational Institute and Corporate Europe Observatory*, 2nd edn.

Ball S, Bell S, 1995 *Environmental Law*, Blackstone Press Ltd. 3rd edn.

Barker K. 2004 *Delivering Stability: Securing our Future Housing Needs*, HMSO, London.

Barlow M, 2008 *Blue Covenant: The Global Water Crisis and the Coming Battle for the Right to Water*, The New Press.

Barlow M, Clarke T, 2002 *Blue Gold: The battle against corporate theft of the world's water*, M and S.

Barnes T, 1996 *Logics of Dislocation: Models, metaphors and meanings of economic space*, Guilford.

Bartlett R.E, 1979, *Developments in sewerage*, Applied Science, Vol 1.

Barty-King H, 1992 *Water: an illustrated history of water supply and wastewater in the United Kingdom*, Quiller Press.

Bauman D.D, Werick W, 1993 "Water Management: why the resistance to change?" *Water Resources Update* 90 (Winter), pp. 3-9.

BBC News, 2006 "Homes forced to get water meters", [on-line] BBC News, UK, http://news.bbc.co.uk/1/hi/england/4759960.stm Accessed 15 November 2006.

BBC News, 2006 "Bringing meters out of the closet", by Mark Kinver, 18 May.

Beder S, 2000 "Hijacking Sustainable Development: A Critique of Corporate Environmentalism", *Chain Reaction* 81, Summer 1999/2000, pp, 8-10.

Beecher Janice A, 1997 "Water Utility Privatization and Regulation: Lessons from the Global Experiment", *Water International* Vol. 22 No. 1.

Berrittella M, 2007 "The economic impact of restricted water supply: A computable general equilibrium analysis", *Water Research* 41(8).

Bessey S.G, 1989 "Problems Encountered in the Hotwells District Small-Scale Metering Trial", *Water and Environment Journal*, 3(6), 579-582.

Bhattacharya, 2003 "European heatwave caused 35,000 deaths", *New Scientist*, 10 October.

Bocking R, 1972 *Canada's Water: for sale?* James Lewis and Samuel.

Bohanna D, 1998 "Water Meters: an incentive to conserve and a signal to the market", *Economic Affairs* 18(2), 10-13.

Bonus H, Niebaum H, 1997, "Benefits and costs of regulating the environment: eight case studies", *Environment and Planning C: Government and Policy* 15(3), 329-346.

Bradford R.B, 2000 "Drought events in Europe", in Vogt, J.V., Somma, F. (eds) *Drought and Drought Mitigation in Europe. Advances in Natural and Technological Hazards Research*, Vol. 14, pp. 7-20. Kluwer.

Brans E, et. al., (eds) 1997 *The Scarcity of Water: Emerging Legal and Policy Responses* Kluwer Law International.

Bressers H, O'Toole L.J, Richardson J, (eds) 1995 *Networks for Water Policy: a comparative perspective*, London: Routledge.

Brouwer R, Pearce D, 2005 *Cost Benefit Analysis and Water Resources*, Edward Elgar, Cheltenham, UK.

Brown R.C, 1991 *Ganesh: studies of an Asian god*, Albany: SUNY Press.

Brownell H, Eaton S, 1975 "The Colorado River Salinity Problem with Mexico." *American Journal of International Law*, 69(255).

Buller H, 1996 "Privatization and Europeanization: The Changing Context of Water Supply in Britain and France", *Journal of Environmental Planning and Management* 39, no. 4, 461.

Bulloch J, Darwish A, 1993 *Water Wars: The coming conflicts in the Middle East*, London: Gollancz.

Butler D, Memon F.A, 2006 *Water Demand Management*. IWA Publishing.

Butler E, Boyfield K, 2002 *Around the World in 80 Ideas 28:The right stuff, user rights and water conservation,* [on-line] Adam Smith Institute, UK. http://www.adamsmith.org/80ideas/idea/28.htm, Accessed 14 November 2006.

Calder I.R, 2005 *Blue Revolution: integrated land and water resource management*, London: Earthscan.

Caponera D.A, 1992 *Principles of Water Law and Administration*, Balkema Publishers.

Carpentier A, Naugues C, Reynaud A, Thomas A, 2004 "Une mesure de l'effet de la délégation sur le prix de l'eau en France", Institut national de la recherche agronomique (INRA) – *Sciences Sociales*, Volume 19, juillet 2004, no. 2.

Cascetta F, 1994 "Application of a portable clamp-on ultrasonic flowmeter in the water industry", *Flow Measurement and Instrumentation* 5(3), 191-194.

Cech T, 2005 *Principles of Water Resources: history, development, management and policy*, Chichester: John Wiley and Sons.

Cellarius B, 1998 "Linking Global Priorities and Local Realities: Nongovernmental Organizations and the Conservation of Nature in Bulgaria," in Krassimira Paskaleva et al., (eds) *Bulgaria in Transition: Environmental Consequences of Political and Economic Transformation* Aldershot, UK: Ashgate Press, pp. 57-82.

Cellarius B, Staddon C, 2002 "Environmental Non-Governmental Organisations, Civil Society and Democratisation in Bulgaria" *Eastern European Politics and Societies*, 16:1.

Chambouleyroun A, 2003 "An incentive mechanism for decentralized water metering decisions", *Water Resources Management* 17(2), 89-111.

Chambouleyroun A, 2004 "Optimal water metering and pricing", *Water Resources Management* 18(4), 305-319.

Chandler W.V, 1984 *The Myth of the TVA: Conservation and development in the Tennessee Valley, 1933-1983*, Ballinger.

Chapagain A.K, 2006 *Globalisation of Water: Opportunities and Threats of Virtual Water Trade*, Taylor and Francis.

Chapman D, (ed.) 1996 *Water Quality Assessments*, 2nd Edn, Chapman and Hall, London.

Chave P, 2001 *The EU Water Framework Directive: An Introduction*, London: IWA Publishing.

Choguill C, Franceys R, Cotton A, 1993 "Planning for water and sanitation", Centre for Development Planning Studies, Loughborough University.

Cimdins P, Druvietis I, Liepa R, Parele E, Urtane L, Urtans A, 1995 "A Latvian cataloque of indicator species of freshwater saprobity", *Proc. of the Latvian Acad. of Sci.*, Section B, No. 1-2 (570-571), 122-133.

Cimdins P, Klavins M, 1998 *Water Quality and Species Diversity of Inland Waters in Latvia*. Verh. Internat. Limnol. 26, 1209-1211.

Clark E, Mondello G, 1999 "Institutional Constraints in Water Management: The French Case", *Water International* Vol. 24 No. 3.

Clarke R, 1993 *Water: The International Crisis*, London: MIT Press.

Clarke R, King J, 2008 *The Atlas of Water: Mapping the Global Crisis in Graphic Facts and Figures*, London: Earthscan.

Collins L, 1991 *European Community Law in the UK*, Butterworths.

Collins R, 1996 *The Waters of the Nile: Hydropolitics and the Jonglei Canal, 1900-1988*. Markus Wiener Publishers, Princeton, NJ.

Commoner B, 1971 *The Closing Circle: Nature, man, technology*, Knopf: New York.

Conca K, 2006 *Governing Water Contentious Transnational Politics and Global Institution Building*, MIT Press

Consumer Council for Water-Midlands, 2006 "Policy on Domestic Metering", draft discussion document.

Cook J, 1989 *Dirty Water*, Unwin Paperbacks

Corr Willbourn Research and Development, 2007 "Deliberative Research into Consumer Views on Fair Charging for the Consumer Council for Water", A Commissioned Research Report, www.corrwillbourn.com.

Cosgrove D, Petts G, (eds) 1990 *Water, Engineering and Landscape*, Belhaven Press.

Costejá M, Font N, Rigol A, Subirats J, 2002 "The evolution of the National Water Regime in Spain", *Euwareness* April 2002.

Cullett P, 2007 *The Sardar Sarovar Dam Project: Selected Documents*, Aldershot: Ashgate.

Dalby S, 1996 "Reading Rio, writing the world: the New York Times and the 'Earth Summit'", *Political Geography* 15 6-7: 593-613.

Darrel Jenerette G, Larsen L, 2006 "A global perspective on changing sustainable urban water supplies", *Global and Planetary Change* 50(3-4), 202-211.

Davis J, Garvey G, Wood M, 1993 *Developing and Managing Community Water Supplies*, Oxfam.

Day G, 2002 "Water Pricing Objectives in England and Wales: institutions and objectives", *Water Science and Technology* 47(6), pp. 33-41.

De Châtel F, 2007 *Water Sheikhs and Dam Builders: Stories of people and water in the Middle East*, Transaction Publishers.

De Villiers M, 2000 *Water*, revised edition, Stoddart.

DEFRA, 2006 *Water Saving Group*, [on-line] Defra, UK. http://www.defra.gov.uk/environment/water/conserve/wsg/pdf/wsg-meteringpaper.pdf, Accessed 14 November 2006.

DEFRA, 2006 *Elliot Morley Urges Action on Wasted Water,*Defra, UK. http://www.defra.gov.uk/news/2006/060308c.htm, Accessed 14 November 2006.

DEFRA, 2006 *Delivering a Sustainable Water supply: Government gives company wider powers to meter,* [on-line] Defra, UK. http://www.defra.gov.uk/news/2006/060301c.htm, Accessed 14 November 2006.

DEFRA, 2006, *Adapting to Climate Change Impacts – A good practice guide for sustainable communities*, [Online] Available from: http://www.defra.gov.uk/science/Project_Data/DocumentLibrary/GA01073/GA01073_4083_FRA.pdf Accessed 19 January 2008.

DEFRA, 2003 Third Consultation Paper on the Implementation of the EC Water Framework Directive (2000/60/EC), London: Department for Environment, Food and Rural Affairs.

DEFRA, 2002 Second Consultation Paper on the Implementation of the EC Water Framework Directive (2000/60/EC), London: Department for Environment, Food and Rural Affairs.

DEFRA, 2001 First Consultation Paper on the Implementation of the EC Water Framework Directive (2000/60/EC), London: Department for Environment, Food and Rural Affairs.

DEFRA, 2000 *Water metering your new rights,* [on-line] Defra, UK. http://www.defra.gov.uk/environment/water/industry/water_metering/rights.htm, Accessed 13 November 2006.

Dellapenna JW, Gupta J (Eds.), 2009 *The Evolution of the Law and Politics of Water*, Springer.

Dellapenna JW, 2001 "The Customary International Law of Transboundary Fresh Waters," *International Journal of Global Environmental Issues* 1:264-305.

Dennis H, 1981 *Water and Power*, Friends of the Earth.

Department for Environment, Food, and Rural Affairs, 2002 *Sewage Treatment in the UK: UK implementation of the EC Urban Waste Water Treatment Directive*, London: DEFRA Publications.

Deutsche W, 2006 "Heat Wave is Drying Up Europe's Water Resources", 2 August.

Diamond B, 1977 Highlights of the Negotiations Leading to the James Bay and Northern Quebec Agreement. Val d'Or: Grand Council of the Cree (of Quebec).

Diamond B, 1990 "Villages of the Damned: The James Bay Agreement Leaves a Trail of Broken Promises," *Arctic Circle*, November/December: pp. 24-34.

Diamond J, 2005 *Collapse: How societies choose to fail or succeed*. Penguin Books. England.

Diamond J, 1998 *Guns, Germs and Steel: A short history of everybody for the last 13,000 years*, Vintage.

Dieperink C, 2000 "Successful International Cooperation in the Rhine Catchment Area", *Water International*, Volume 25, Number 3.

Donahue J.M, Rose B, (eds) 1997 *Water, Culture and Power: Local struggles in a global context*, Island Press.

Dovey W.J, Rogers D.V, 1993 "The Effect of Leakage Control and Domestic Metering on Water Consumption in the Isle of Wight", *Water and Environment Journal* 7(2), 156-160.

Downing R.A, 1998 *Groundwater. Our Hidden Asset*, British Geological Survey.

Doyle T, McEachern, 1998 *Environment and Politics*, London: Routledge.

Drozdov S, 2002 "Use of Water Consumption Metering as a Tariff Policy Tool: Moldova's Experience".

Drury I, 2007 "Meters could add £200 to the family water bill", *Daily Mail*, 15 October.

Easter K.W, Renwick M.E, 2004 *Economics of Water Resources: Institutions, instruments and policies for managing scarcity*, Aldershot: Ashgate.

Edwards K, Martin L, 1995 "A Methodology for Surveying Domestic Water Consumption" *Water and Environment Journal* 9(5), 477-488.

El Zain M, 2007 *Environmental Scarcity, Hydropolitics, and the Nile: Population Concentration, Water Scarcity and the Changing Domestic and Foreign Politics of the Sudan*, Shaker Publishing.

Engleman R, Leroy P, 1992 *Sustaining Water: Population and the future of renewable water supply*, Population Action International.

Environment Agency, 2008 *Managing Water Abstraction*: interim update, HMSO.

Environment Agency, 2007 *Water Resources and Abstraction*, [Online] Available from: http://www.environment-agency.gov.uk/yourenv/eff/1190084/water/213872/609264/?version=1&lang=_e, Accessed on: 13 November 2007.

Environment Agency, 2006 *The Test and Itchen Catchment Abstraction Management Strategy* [online] www.publications.environment-agency.gov.uk.

Environment Agency, 2002 *The Water Framework Directive: Guiding principles on the technical requirements*. Bristol: Environment Agency.

Environment Agency, 1998 *Assessing Water Quality – General Quality Assessment Scheme for Biology*, Bristol: Environment Agency.

Epple R, 1996 "Important Victory on the Elbe River" *World Rivers Review*, 11(5).

European Commission, 2000 *Directive 2000/60/EC of the European Parliament and of the Council of 23rd October 2000 establishing a framework for Community action in the field of water policy*, Official Journal 22 December 2000 L 327/1, Brussels: European Commission.

European Commission, 1998, *Towards Sustainable Development: Guidelines for water resources development co-operation, a strategic approach*, Brussels: Commission of the European Communities:DGVIII, ISBN 92-828-4454-4.

European Environment Agency, 2009 "Water Resources Across Europe: confronting water scarcity and climate change", EEA, #2/2009.

Fahim H, 1981 *Dams, People and Development: the High Aswan Experience*, Pergamon, Oxford.

Faugli P.E, 1994 "Watercourse Management in Norway", *Norsk Geografisk Tidsskrift* 48(1-2), pp. 75-79.

Finger M, 2004 *Water Privatisation: Trans-national corporations and the re-regulation of the water industry*, Oxford: Blackwell.

Fischer W.F, (ed.) 1995 *Towards Sustainable Development? Struggling over India's Narmada River*, ME Sharpe, Armonk, NJ.

Fisher D.E, 2000 *Water Law.* Law Book Company.

Fishlock T, 1972 *Wales and the Welsh*, London: Cassell and Company.

Fitzmaurice J, 1998 *Damming the Danube: Gabcikovo/Nagymaros and Post-communist Politics in Europe*, Westview Press.

Foster H, Beattie B, 1979 "Urban Residential Demand for Water in the United States", *Land Economics* 55(1), 43-58.

Foundation for Water Research, 2004 *The EC Water Framework Directive – An Introductory Guide*, An FWR Guide, August.

Gabcíkovo-Nagymoros Case (Hungary v. Slovakia), 1997 ICJ No. 92.

Gadbury D, Hall M.J, 1989 "Metering trials for water supply", *Journal – Institution of Water and Environmental Management* 3(2), 182-187.

Gale I, Williams A, Gaus I, Jones H, 2002 *ASR-UK: elucidating the hydrogeological issues associated with aquifer storage and recovery in the UK*, UKWIR Report Ref. No. 02/WR/09/2.

Gaudin S, 2006 "Effect of price information on residential water demand", *Applied Economics* #38, 383-393.

Gaufin A.R, Tarzwell C.M, 1952 "Aquatic invertebrates as indic ators of stream pollution". Public Health Reports 67: 57-64.

Gleick P.H, 1999a "The Human Right to Water", *Water Policy* 1(5) 487-503.

Gleick P.H, 1999b *The World's Water: The Biennial Report on Freshwater Resources*, Island Press.

Gleick P.H, 1993 *Water in Crisis: A guide to the world's fresh water resources*, Oxford University Press.

Gleick P.H, Wolff, Chalecki G, Reyes E.L, 2002 *The New Economy of Water: The Risks and Benefits of Globalization and Privatization of Fresh Water*, Pacific Institute for Studies in Development, Environment, Security.

Glennie E.B, et al., 2002 "Phosphates and Alternative Detergent Builders: final report", EU Environment Directorate.

Glennon R, 2002 *Water Follies: Groundwater pumping and the fate of America's fresh waters*, Island Press.

Goldsmith E, Hildyard N, 1992 *The Social and Environmental Effects of Large Dams*, three volumes, Wadebridge Ecological Centre.

Goodman A.S, Edwards K.A, 1992 "Integrated Water Resources Planning", *Natural Resources Forum* 17(1), pp. 33-42.

Gopalakrishnan C, Okada N, (eds) 2008 *Water and Disasters*, London: Routledge.

Goudie A, 1993 *The Human Impact on the Natural Environment*, Blackwell.

Graffy C.P, 1998 "Water, Water, Everywhere, Nor Any Drop to Drink: The Urgency Of Transnational Solutions To International Riparian Disputes", *Geo. Int'l Envtl. L. Rev.* 10, 399.

Grant W, 2000 *The Effectiveness of European Union Environmental Policy*, Basingstoke: Macmillan.

Green C, 2003, *Handbook of Water Economics: Principles and practice*, Chichester: Wiley.

Griffin R.C, 2006 *Water Resource Economics: The analysis of scarcity, policies, and projects*, London: MIT Press.

Groombridge B, Jenkins M.D, 1996 "Assessing Biodiversity Status and Sustainability", *World Conservation Monitoring Centre Biodiversity Series* No. 5. pp. 104.

Guy S, Marvin S, Moss T, 2001 *Urban Infrastructure in Transition*, London: Earthscan.

Hagen L.H, 2000 "Handing Water Shortage in Oslo", *Water Supply* 18(1-2), 381-384.

Haigh N, 1989 *EEC Environmental Policy and Britain*, Longman (2nd edn).

Halcrow, 2007 *Corby Water Cycle Strategy Phase 2 – Detailed Strategy Summary Report* February.

Hall D, 1999 "The Water Multinationals", presented at: PSI conference on water industry, Bulgaria, October 1999; OTV conference on water industry in Germany, Essen, October 1999.

Hall D, Lobina E, 2004 "Private and Public Interests in Water and Energy", *Natural Resources Forum* 28, pp. 268-277.

Hall M.J, Hooper B.D, Postle S.M, 1988 "Domestic per capita water consumption in South West England". *Journal – Institution of Water and Environmental Management* 2(6), 626-631.

Halliday S, 2001 *The Great Stink of London: Sir Joseph Bazalgette and the Cleansing of the Victorian Metropolis*, Sutton Publishing.

Hanæus, Hellström, Johansson, 1997 "A study of a urine separation system in an ecological village in northern Sweden", *Water Science and Technology*, Vol 35(9).

Hanley N, 2001 *Introduction to Environmental Economics*, Oxford: Oxford University Press.

Hardin G, 1998 "Extensions of "The Tragedy of the Commons"", *Science* 280, 682-683.

Hardin G, 1968 "The Tragedy of the Commons", *Science*, 162, pp. 1243-1248.

Harries T, 2007 "Householder Responses to Flood Risk: The consequences of the search for ontological security." PhD Thesis, Flood Hazard Research Centre, University of Middlesex.

Harvey D, 1990 *The Condition of Postmodernity: An enquiry into the origins of cultural change,* Blackwell.

Hassan, J 1998 *A history of water in modern England and Wales*, Manchester University Press

Hauck G.F.W, Novak R.A, 1987 "Interaction between flow and incrustation in the Roman aquaduct of Nimes" *Journal of Hydraulic Engineering* 113(2), 141-157.

Havlicek R, 2005 "Learning from Privatisation of Water Services in Trencin, Slovakia", in Balanyá et al. (eds) *Reclaiming Public Water: Achievements, Struggles and Visions from Around the World, Transnational Institute and Corporate Europe Observatory*, 2nd edition, pp. 201-212.

Heritier A, 2001 "Market integration and social cohesion: the politics of public services in European regulation", *Journal of European Public Policy* 8:5 October: 825-852.

Hester R.E, Harrison R.M, (eds) 1995 *Waste Treatment and Disposal*, Cambridge: Royal Society of Chemistry.

Hirschleifer A, 1960 *Water Supply: Economics, Technology and Policy*, Linden Publishers New York.

Hobson S, 2007 "Water Leak figures trump estimates", *Utility Week*, 19 October.

Hochstrat R, Melin T, Wintgens T, 2008 "Development of Integrated Water Reuse Strategies", *Desalination*, Issue 218, p. 208-217.

Hodge T, 2001 *Roman Aquaducts and Water Supply*, Duckworth.

Hoekstra A, Chapagain A, 2007 "Water Footprints of Nations: water use by people as a function of their consumption patterns", *Water Resource Management*, 21(1).

Holland A.C, 2005 *The Water Business: Corporations vs people*, London: Zed Books.

Horn R, 2001 *Dictionary of Water*, Gottingen: Steidl.

House of Commons Environment, Food and Rural Affairs Committee, 2008 *Implementation of the Nitrates Directive in England*, HMSO.

Howe C.W, 1971 "Benefit-Cost Analysis for Water System Planning", *Water Resources Monograph No 2, American Geophysical Union*, Washington DC.

Howsam P, Carter R, 1996 *Water policy*, London: E and FN Spon.

Howsan P, Carter R, 1996 *Allocation and Management in Practice*, London: E and FN Spon.

Hulme M, Jenkins G, 1998 "Climate change scenarios for the United Kingdom". Scientific Report. Technical Report No. 1. UK Climate Impacts Programme.

Hungarian Central Statistical Office, 2004 *Public Utilities Report.*

International Conference on Water and the Environment (ICWE), 1992 The Dublin Principles. Available in full at http://www.wmo.ch/web/homs/icwedece.html.

International Law Association, 2004 ("ILA 2004"), "The Berlin Rules on Water Resources," in Report of the Seventy-First Conference (Berlin). International Law Association, London, UK.

International Law Association, 1966 ("ILA 1966"), "The Helsinki Rules on the Uses of the Waters of International Rivers," in Report of the Fifty-Second Conference (Helsinki). International Law Association, London, UK.

IPCC, 2007 *Climate Change 2007: The Physical Science Basis. Summary for Policymakers.*

Jarvie J, 2007 "Atlanta Water Use is Called Shortsighted", *Los Angeles Times*, 4 November, p. A-15.

Jefferson B, Laine A, Parsons S, Stephenson T, Judds S, 1999 "Technologies for Domestic Waste Water Recycling", *Urban Water* 1, pp. 285-292.

Jenkins J.O, 2006 "Water Metering: in search of a more critical debate", unpublished paper, CCW-Thames Region.

Jonsdottir H, Eliasson J, Madsen H, 2005 "Assessment of serious water shortage in the Icelandic water resource system", *Physics and Chemistry of the Earth, Parts A/B/C* (6-7) 420-425.

Kaczmarek B, 1990 "The Use of Economic Measures for Water Management in France" *Natural Resources Forum* 14(4) 312-318.

Kaika M, Page B, 2002 "The making of the EU Water Framework Directive: Drifting choreographies of governance and the effectiveness of environmental lobbying", (http://www.geog.ox.ac.uk/research/wpapers/economic/wpg02-12.pdf).

Kaika M, 2005 *City of Flows: Modernity, Nature and the City*, London: Routledge.

Kallis G, Butler D, 2001 "The EU Water Framework Directive: Measures and Implications", *Water Policy* 3(3), 125-142.

Kallis G, Nijkamp P, 2000 "Evolution of EU Water Policy: A critical assessment and a hopeful perspective", *Journal of Environmental Law and Policy* 3/2000, 301-335.

Kalotay K, Hunya G, 2000 "Privatisation and FDI in Central and Eastern Europe", *Transnational Corporations* 9(1): 39-66.

Karkkainen B, 2004 "Post-sovereign environmental governance" *Global Environmental Politics* 4(1), 72-96.

Katko T.S, 1992 "Evolution of Consumer-managed Water Cooperatives in Finland, with Implications for Developing Countries", *Water International* Vol. 17 No. 1.

Kay M.F, Smith L, 1997 *Water: Economics, management and demand*, London: E and FN Spon.

Kay S.B, 1998 "Metering for Demand Management: The Cambridge experience", *Journal of the Chartered Institution of Water and Environmental Management* 12(1), 1-5.

Kindler J, 1992 "Water Management in Central and Eastern Europe", *Ecodecision*, September, pp. 48-50.

King D, 2007 "A Measure of Change", *Utility Week*, 26 October.

King D, 2006 "Managing Water Resources in England and Wales, a lecture by David King", Director of Water Resources, Environment Agency, 6 June 2006 at the Foundation for Science and Technology, London, UK.

Kinnersley D, 1994 *Coming Clean: The politics of water and the environment*, London: Penguin.

Kinnersley D, 1988 *Troubled Water*, London: Hilary Shipman.

Kirby C, (ed.) 1994 *Integrated River Basin Management*, Chichester: Wiley.

Kjaerland F, 2007 "A Real Option Analysis of Investments in Hydropower: The case of Norway" *Energy Policy* 35(11), 5901-5908.

Kliot N, Shmueli D, 2001 "Development of Institutional Frameworks for the Management of Transboundary Water Resources", *International Journal of Global Environmental Issues*, 306.

Knight G, et al., 1995 "The Emerging Water Crisis in Bulgaria," *Geojournal* volume 35(4), pp. 415-423.

Kompetenzzentrum, 2008 *Water Industry in Berlin-Brandenburg, Report 2008*, Kompetenzzentrum Wasser Berlin GmbH.

Konig K, 2007 "Rainwater Management: examples from Germany" paper presented at the International Water Conference, 12-14 September 2007, Berlin, Germany.

Kravcik M, et al., 2002 *Blue Alternative for Blue Rivers*.

Krinner W, et al., 1999 *Sustainable Water Use In Europe, Part 1: Sectoral use of water*, Luxembourg: Office for Official Publications of the European Communities.

Landre B.K, Knuth B.A, 1993 "Success of Citizen Advisory Committees in Consensus-based Water Resources Planning in the Great Lakes Basin", *Society and Natural Resources* 6(3), pp. 229-257.

Larssen, T, et al., 2006 "Acid Rain in China", *Environmental Science and Technology*, 40:2, pp. 418-425.

LaRusic I, et al., 1979 *Negotiating a Way of Life: Initial Cree Experiences with the Administrative Structure Arising from the James Bay Agreement*. Ottawa: Canada, Department Of Indian and Northern Affairs, Policy Research and Evaluation Group.

Laver M, 1984 "The Politics of Inner Space: tragedies of three commons", *European Journal of Political Research* 12, pp. 59-71.

Lean G, Hinrichson D, 1992 *Atlas of the World's Environment*, HarperCollins.

Lecornu J, 1999 "Multipurpose Dams: The future of dams", address at the Polish National Committee on Large Dams VIII Conference of Technical Inspection of Dams, 20-23 June 1999 Zakopane-Koscielisko, Poland.

Leonard J, Crouzet P, 1998 "Lakes and Reservoirs in the EEA Area", European Environmental Agency.

Levi P, 1995 *The Periodic Table*, Penguin Books.

Lewis S, 1996 *The Sierra Club Guide to Safe Drinking Water*, Sierra Club Books.

Lloyd-Hughes B, Saunders M.A, 2002 "A Drought Climatology for Europe" *International Journal of Climatology* 22: 1571-1592.

Lobina D, Hall D, 2000 "Public Sector Alternatives to Water Supply and Sewerage Privatization: Case Study", *International Journal of Water Resources Development* Vol. 16 Issue 1, p. 35-56.

Lobina E, 2008 WaterTime National Context Report – Italy.

Lobina E, 2005 "Problems with Private Water Concessions: A review of experiences and analysis of dynamics", *International Journal of Water Resources Development* 21(1): 55-87.

Lomborg B, 1998 *The Skeptical Environmentalist: Measuring the real state of the world*, Cambridge.

Lorrain D, Stoker G, (eds) 1997 *The Privatization of Urban Services in Europe*, Pinter.

Lowi M, 1995 *Water and Power: The Politics of a Scarce Resource in the Jordan River Basin*, Cambridge University Press.

Lueck Andreas H, 1999 "Transboundary Water Resources Management and the European Union's (EU) Water Policy", *Water International* Vol. 24 No. 1.

Luke T, 1995 "Sustainable Development as a power/knowledge system: the problem of governmentality" in F. Fischer and M. Black (eds) *Greening Environmental Policy: the politics of a sustainable future,* Paul Chapman Publishing, 21-32.

Lund J.R, 1988 "Metering utility services: evaluation and maintenance", *Water Resources Research* 24(6), 802-816.

Lundberg D.C, 1992 *The Beginnings of Western Science*, University of Chicago Press.

Lundqvist J, 2000 "Rules and Roles in Water Policy and Management" *Water International* Volume 25, Number 2.

MacDonnell L.J, et al., 1995 "The Law of the Colorado River: Coping with Severe Sustained Drought" *Water Resources Bulletin* 31, pp. 825-836.

MacLeod, 1979 "The effect of metering on urban water consumption (Durban, South Africa)", *Municipal Engineer (Johannesburg)* 10(3), 23-26.

Mahmoud A, Atef H, Lacirignola C, 1995 "Water Crisis in the Mediterranean: Agricultural Water Demand Management", *Water International,* Vol. 20 No. 4.

Malik L.K, Koronkevich N, Zaitseva I.S, Barabanova E.A, 2000 Briefing Paper on Development of Dams in the Russian Federation and other NIS Countries, Case Study prepared as an input to the World Commission on Dams, Cape Town, http://www.dams.org/.

Maloney W.A, Richardson J, 1995 "Water Policy-making in England and Wales: Policy communities under pressure?" in H. Bressers, L. O'Toole Jnr, J. Richardson (eds) *Networks for Water Policy: A comparative perspective*, Frank Cass, London.

Manner E.J, Metsalampi V.M, (eds) 1988 *The Work of the International Law Association on the Law of International Water Resources.*

Martin W.E, Ingram H.M, Laney N.K, Griffin A.H, 1984 *Saving Water in a Desert City* Washington, DC: Resources for the Future.

Marvin S, Chappells H, Guy S, 1999 "Pathways of Smart Metering Development: Shaping environmental innovation", *Computers, Environment and Urban Systems* 23(2), 109-126.

Masciopinto C, 1999 "Barbiero Giulia, Benedini Marcello, A Large Scale Study for Drinking Water Requirements in the Po Basin (Italy)", *Water International* Vol. 24 No. 3.

McCaffrey S, 2001 *The Law of International Watercourses: Non-navigational uses*, Oxford University Press.

McCaffrey S, 1992 "A Human Right to Water: Domestic and international implications", *Georgetown International Environmental Law Review* 5(1), 1-238.

McCaleb J.W, 1999 "Stewardship Of Public Lands and Cultural Resources in the Tennessee Valley: A Critique of the Tennessee Valley Authority", *Re Communes: Vermont's Journal of the Environment.*

McCormick J, 1997 *Acid Earth: The Global Threat of Acid Pollution*, London: Earthscan.

McCully P, 1996 *Silenced Rivers: The ecology and politics of large dams*, London: Zed Books.

McCutcheon S, 1991 *Electric Rivers, The Story of the James Bay Project.* Montreal: Black Rose Books.

McDonald A, Kay D, 1988 *Water Resources*, Harlow: Longmans.

Medd W, Shove E, 2006a *Traces of water workshop report 5: Imagining the future,* [on-line] Lancaster University, UK. http://www.lec.lancs.ac.uk/cswm/download.TWW5_Report.pdf, Accessed 20 November 2006.

Medd W, Shove E, 2006b *Traces of Water Workshop Report 4; Water stresses and the consumer,* [on-line] Lancaster University, UK. http://www.lec.lancs.ac.uk/cswm/download/TWW4_Report.pdf, Accessed 20 November 2006.

Medd W, Shove E, 2005a *Traces of Water Workshop Report 2: Water practices in everyday life,* [on-line] Lancaster University, UK, http://www.lec.lancs.ac.uk/cswm/download/tww2_report1.pdf, Accessed 20 November 2006.

Medd W, Shove E, 2005b *Traces of Water Workshop Report 1: Perspectives on the water consumer,* [on-line] Lancaster University, UK. http://www.lec.lancs.ac.uk/cswm/download/tww1_report1.pdf, Accessed 20 November 2006.

Meinck F, Mohle H, 1963 *Dictionary of Water and Sewage Engineering*, London: Elsevier.

Mercer D, Christensen L, Buxton M, 2007 "Squandering the Future – Climate change, policy failure and the water crisis in Australia", *Futures* 39(2-3), 272-287.

Merrett S, 1999 "The Regional Water Balance Statement: A New Tool for Water Resources Planning", *Water International* Vol. 24 No. 3.

Merrett S, 1996 *Introduction to the Economics Water Resources*, London: UCL Press.

Merritt W.S, Alila Y, Barton M, Taylor B, Cohen S, Neilsen D, 2006 "Hydrologic response to scenarios of climate change in sub watersheds of the Okanagan basin, British Columbia", *Journal of Hydrology* 326(1-4), 79-108.

Micklin P.P, 1977 "NAWAPA and Two Siberian Water-Diversion Proposals: A Geographical Comparison and Appraisal", *Soviet Geography: Review and Translation* Vol. 18, No. 2, pp. 81-99.

Mitchell B, 1997 *Resource and Environmental Management*, Longman.

Mitchell B, 1990 *Integrated Water Management: International experiences and perspectives*, Bellhaven.

Mitchell B, Shrubsole D, 1994 *Canadian Water Management: Visions for sustainability*, Cambridge, Ont: Canadian Water Resources Assn.

Moore C.W, 1994 *Water and Architecture*, Harry N Abrams Inc.

Murphy I, 1997 *The Danube: A River Basin in Transition*.

Mylopoulos Y.A, Kolokytha E.G, 2008 "Integrated water management in shared water resources: The EU Water Framework Directive implementation in Greece", *Physics and Chemistry of the Earth, Parts A/B/C*, 33(5): 347-353.

Nacheva V, 2002 "Kuwait Eyes Bulgarian Water Project", *Sofia Echo*, 17 May.

National Academies, 2008 *Understanding and Responding to Climate Change*, NAS: Washington.

National Research Council, 2002 *Privatisation of Water Services in the United States*, National Academy Press.

National Rivers Authority, 1994 *Water: Nature's precious resource – An environmentally sustainable water resources development strategy for England and Wales*, HMSO.

National Water Metering Trails Working Group, 1993 *National Metering Trials Final Report*, Water Services Association.

Newson M, 2008 *Land, Water and Development*, third edition, London: Routledge.

Newson M, 1992 *Land, Water and Development*, first edition, London: Routledge.

Niemczynowicz J, 1996 "Megacities from a water perspective", *Water International* 21(4), pp. 198-205.

No Author, 1955 *City of Birmingham Waterworks: A Short History of the Development of the Undertaking, with a Description of the Existing Works and Sources of Supply*. Birmingham: City of Birmingham, Water Committee.

Noel J.-C, 1990 "Water Policy in France: The River Basin Agencies", *Water International*, Vol. 15 No. 2.

Norris H, 2001 *The Great Thirst: Californians and water: a history*, London: University of California Press.

Novotny V, Somlyody L, (eds) 1995 *Remediation and Management of Degraded River Basins: With emphasis on Central and Eastern Europe*, Berlin: Springer.

O'Riordan T, Cameron J, (eds) 1994 *Interpreting the Precautionary Principle*, Cameron May.

O'Riordan T, Voisey H, 1997 "Sustainable development in Western Europe: Coming to terms with Agenda 21", *Environmental Politics* 6(1), 1-177.

O'Tuathail G.O, 1994 "The Critical Reading/writing of Geopolitics: Re-reading/ writing Wittfogel, Bowman and Lacoste", *Progress in Human Geography* 18(3) pp. 313-332.

ODA, 1993 *Priorities for Water Resources Allocation and Management*, London, ODA. (Copies from The Distribution Office, NRI, Chatham Maritime, ME4 4TB.)

ODPM, 2003 *Sustainable Communities: Building for the Future*, TSO, London.

OECD, 2003 *Social Issues in the Provision and Pricing of Water Services*. OECD: Paris.

OECD, 1999 *The Price of Water: Trends in OECD Countries*. OECD: Paris.

OECD, 1989 *Water Resources Management: Integrated Policies*, OECD, Paris. OECD 2 rue Andrea Pascal, 75775, Paris; HMSO, PO Box 276, London SW8 5DT.

Ohlsson L, 1995 *Hydropolitics: Conflicts over water as a development constraint*, London: Zed Books.

Olivera O, 2004 *Cochabamba: Water war in Bolivia*, South End Press.

Opinion Leader Research, 2006 "Using Water Wisely: a deliberative consultation", research conducted for the Consumer Council for Water, UK.

Orwin A, 1999 "The Privatization of Water and Wastewater Utilities: An International Survey", *Environmental Probe* [on-line], http://www.environmentprobe.org/enviroprobe/pubs/ev542.html, accessed 18 July 2002.

Ostrom E, et al., 2000 "Reformulating the Commons," *Swiss Political Science Review*, Vol. 6, No. 1, pp. 29-52.

Ostrom E, 1999 "Revisiting the Commons: Local Lessons, Global Challenges", *Science* 284, 278-282.

Ostrom E, 1990 *Governing the Commons: The Evolution of Institutions for Collective Action*, Cambridge: Cambridge University Press.

Paisley R, 2002 "Adversaries into Partners: International water law and the equitable sharing of downstream benefits", *Melbourne Journal of International Law* 3, 280-300.

Parker S, 2003 *Thirsty world*, Oxford: Heinemann.

Pavlinek P, Pickles J, 2004 *Environmental Transitions: Transformation and ecological defence in central and eastern Europe*, London: Routledge.

Pearce D, and Turner R.K, 1990 *Economics of Natural Resource and Environmental Management*, Harvester Wheatsheaf Publications.

Pearce F, 2006 *When the Rivers Run Dry: What happens when our water runs out?* London: Eden.

Penkova N.V, Shiklomanov I.A, 1998 "Methodological Principles for Assessment and Prediction of Water Use and Water Availability in the World", *Water: A looming crisis?* (H. Zebidi, ed.). Proc. International Conference on World Water Resources at the Beginning of 21st Century (UNESCO, Paris, 3-6 June 1998). IHP-Y Technical Documents in Hydrology, 18, 25-36.

Perkins P, 2007 "Solving Canberra's Water Woes", *Canberra Times*, 26 October 2007.

Pheko M, 2007 "Policies perpetuate the invisible racism rampant in our country" *Sunday Times* (South Africa), 14 October 2007.

Pint E.M, 1999 "Household Responses to Increased Water Rates During the California Drought", *Land Economics* 75(2), 246-266.

Pitt M, 2007 "Learning lessons from the 2007 floods" UK Cabinet Office.

Porter R.C, 1996 *The Economics of Water and Waste: A Case Study of Jakarta, Indonesia,* Aldershot: Ashgate.

Postel S, 2005 "From the Headwaters to the Sea: The critical need to protect freshwater ecosystems" *Environment* 47(10), 8-22.

Postel S, 1992 *Lost Oasis: Facing H_2O scarcity*, WW Norton.

Pringle L, 1982 *Water: The next great resource battle*, Macmillan Publishing.

Pryde, 1987 "The Soviet Approach to Environmental Impact Analysis," in Singleton F, (ed.) *Environmental Problems in the Soviet Union and Eastern Europe*, Lynne Reinner: Boulder, CO., pp. 43-58.

PSPRU Paper, 1997 "Water Concessions: Ownership, Control and Regulation" (August).

Public Citizen, 2002 *Profit Streams: The World Bank and Greedy Global Water Companies*, Public Citizen. [online] http://www.citizen.org/documents/ ProfitStreams-World%20Bank.pdf.

Punys P, Pelikan B, 2007 "Review of small hydropower in the new Member States and Candidate Countries in the context of the enlarged European Union", *Renewable and Sustainable Energy Reviews*, 11(7) 1321-1360.

Ravasteijn W, Kroesen J, 2007 "Tensions in water management: Dutch tradition and European policy", *Water Science and Technology*, 56(4): 105-111.

Reisner M, 1993 *Cadillac Desert: The American west and its disappearing water*, Penguin Books.

Richardson B, 1979 *Strangers Devour The Land.* Vancouver: Douglas and McIntyre.

Richardson J.J, Maloney W.A, Rudig W, 1992 *Privatising Water*, Strathclyde Papers on Government #80.

Robson A, Howsam P, 2006 "Domestic Water Metering – Is the law adequate?" *Journal of Water Law* 17(2), pp. 65-70.

Rogers P, 2008 "Facing the Freshwater Crisis", *Scientific American* August, pp. 46-53.

Roth D, Boelens R, 2005 *Liquid Relations: Contested water rights and legal complexity*, New London: Rutgers University Press.

Rothfeder J, 2002 *Every Drop for Sale: The World's Desperate Fight over Water*, Jeremy P Tarcher.

Royte E, 2008 *Bottlemania: How water went on sale and why we bought it*, Bloomsbury.

Salisbury R.F, 1986 *A Homeland for the Cree. Regional Development in James Bay, 1971-1981*. Montreal: McGill-Queen's University Press.

Salman A.M.S, McInerney-Lankford S, 2004 *The Human Right to Water: Legal and Policy Dimensions*, World Bank; Washington, DC.

Sampson S, 1996 "The Social Life of Projects: Importing Civil Society to Albania," in C. Hann and E. Dunn (eds) *Civil Society: Challenging Western Models* (London: Routledge, 1996), pp. 21-142.

Samson P, Charrier B, 1997 *International Freshwater Conflict: Issues and Prevention Strategies*, Green Cross International.

Sanderson M.L, 1994 "Domestic water metering technology", *Flow Measurement and Instrumentation* 5(2), 107-113.

Scheidleder A, et al., 1999 *Groundwater Quality and Quantity in Europe*, Office for Official Publications of the European Communities.

Scott C.H, 1989 "Ideology of Reciprocity Between the James Bay Cree and the Whiteman State," in P. Skalnik (ed.) *Outwitting the State*. New Brunswick, N.J.: Transaction Publishers.

Seager J, 1993 "Statutory Water Quality Objectives and River Water Quality", *Water and Environment Journal* 7 (5), 556-562.

Segerfeldt F, 2005 *Water for Sale: How Business and the Market Can Resolve the World's Water Crisis*, Cato Institute.

Selby J, 2003 *Water, Power and Politics in the Middle East*, London: I.B. Tauris.

Serageldin I, 1995 *Toward Sustainable Management of Water Resources*, Washington DC: The World Bank.

Serageldin I, 1995 "Water Resources Management: A new policy for a sustainable future", *Water Resources Development*, 2(3).

Severn Trent Water PLC, 2006 "Customer Priorities and Willingness to Pay", final report on research conducted for Severn Trent by Accent, London, UK.

Sherk G.W, Wouters P, Rochford S, 2002 *Water Wars in the Near Future? Reconciling Competing Claims for the World's Diminishing Freshwater Resources – The Challenge of the Next Millennium*, The CEPMLP On-Line Journal, Vol. 3, No.2.

Shiklomanov I.A, Rodda J.C, 2004 *World Water Resources at the Beginning of the Twenty-First Century*, Cambridge University Press.

Shiva V, 2002 *Water Wars: Privatization, pollution and profit*, London: Pluto.

Shove E, Watson M, Hand M, Ingram J, 2007 *The Design of Everyday Life*, Berg Publishers.

Shove E, 2003 *Comfort, Cleanliness and Convenience: The social organization of normality*, Oxford: Berg.

Simonson J, 2003 *The Global Water Crisis: NGO and Civil Society Perspectives*, Centre for Applied Studies in International Negotiation, Geneva.

Sloep P.B, Blowers A, 1996 *Environmental Policy in an International context*, Edward Arnold.

Smith A.L, Rogers D.V, 1990 "Isle of Wight water metering trial", *Journal – Institution of Water and Environmental Management* 4(5), 403-409.

Smith K, 1972 *Water in Britain*, Macmillan.

Smith M.G, de March C, Jolma A, 1999 "Paddling Enforceable Approaches Upstream to EU Standards: Water quality management and policy implementation in Central and Eastern Europe", *Water Science and Technology*, 49(10), pp. 73-79.

Solanes M, Gonzalez-Villarreal F, 1999 *The Dublin Principles for Water as Reflected in a Comparative Assessment of Institutional and Legal Arrangements for Integrated Water Resources Management*, Technical Advisory Committee, Global Water Partnership (June).

Solley W.B, Pierce R.B, Perlman H.A, 1998 *Estimated Use of Water in the United States in 1995*, U.S. Department of the Interior. US Geological Society. Denver, CO: USGS.

Squatriti P, (ed.) 2001 *Managing Water in Medieval Europe: Technology and Resources*, Brill.

Staddon C, 1998 "Democratisation and the Politics of Water Management in Bulgaria," J. Pickles, A. Smith (eds) *Theorising Transition: The political economy of post-communist transformations*, London: Routledge, pp. 347-372.

Staddon C, 1996 *Democratisation, Environmental Management and the Production of New Political Geographies in Bulgaria: a case study of the 1994-95 Sofiya Water Crisis*, unpublished PhD Thesis, Department of Geography, University of Kentucky.

Stahl H, Tallaksen K, Demuth L.M, 2001 "Have streamflow droughts in Europe become more severe or frequent?", *International Journal of Climatology*, 21: 3, 317-333.

Stanners D, Bourdeau P, 1995 *Europe's Environment: The Dobris Assessment*, European Environmental Agency.

Sternberg R, 2008 "Hydropower: Dimensions of social and environmental coexistence", *Renewable and Sustainable Energy Reviews*, 12(6) 1588-1621.

Swyngedouw E, 2005 *Social Power and the Urbanization of Water: Flows of Power* (Oxford Geographical and Environmental Studies), Oxford University Press.

Swyngedouw E, 2004 "Modernity and Hybridity: Nature, Regeneracionismo, and the Production of the Spanish Waterscape, 1890-1930", in T. Barnes, J. Peck, E. Sheppard, and A. Tickell (eds) *Reading Economic Geography*. Basil Blackwell, Oxford, pp. 189-204. (reprinted from Annals of the Association of American Geographers, 89(3), pp. 443-465).

Szulinska J, 2002 "Water systems in Poland – the safety aspects". http://www. enpc.fr/cereve/HomePages/thevenot/www-jes-2002/Szulinska-Paper-2002. pdf.

Tainter J.A, 1990 *The Collapse of Complex Societies*. Cambridge: Cambridge University Press.

Taylor G, 1999 *State Regulation and the Politics of Public Service: The case of the water industry*, Mansell.

Taylor P, 2007 "The water industry's response to climate change" address given at Water UK Climate Change Conference, Wednesday 26 September 2007, London.

Tebbutt T.H.Y, 1998 *Principles of Water Quality Control* (5th Edition), Butterworth Heinemann, London.

Teclaff L.A, 1985 *Water Law in Historical Perspective*, Buffalo, NY: William S. Hein Co.

ten Brinke W, 2005 *The Dutch Rhine: A restrained river*, Veen Magazines BV.

Thomas A, Millard A, 1994 "Budget metering in the water sector", *Flow Measurement and Instrumentation* 5(2), 143-143.

Thomas C, Ingrams K, Dhayatker K, Searle C, 2008 "Assessing the impact of Sea Level Rise (SLR) and an increase in storm surge events (SSE) on a section of coastline; Sand Point to Brean Down, Somerset", unpublished research report, Department of Geography and Environmental Management, University of the West of England, Bristol.

Thomas E, 2007 *Capel Celyn, Ten Years of Destruction: 1955-1965*, Cyhoeddiadau Barddas and Gwynedd Council.

Thomas G, Wilson S, 2007 *The Gloucestershire Floods 2007*, The History Press.

Thukral E.G, (ed.) 1992 *Big Dams, Displaced People: Rivers of sorrow, rivers of change*, Sage Publications.

Trottier J, (ed.) 2007 *Politics of Water: A survey*, London: Routledge.

Tuan Y, 1968 *The Hydrologic Cycle and the Wisdom of God: A theme in geoteleology*, University of Toronto Press.

United Nations, 2008 *The Millennium Development Goals Report*. Geneva.

United Nations Food and Agriculture Organization, 1998 *Sources of International Water Law*, Legislative Study 65.

United Nations, 1983 *Mar del Plata Action Plan (Water Resources development in Developing Countries)*.

United Nations, 1977 Report of the United Nations Water Conference, Mar del Plata. 14-25 March 1997. No. E.77.II.A.12, United Nations Publications, New York.

United Utilities, 2001/2 *Annual Reports 2001, 2002*.

Van der Brugge R, Rotmans J, 2007 "Towards transition management of European water resources", *Water Resources Management*, 21(1).

van der Merwe L.H, 2003 "Metering in demand: Uncovering prepayment". *Water Sewage and Effluent* 23(5), 30-33.

Van Vliet B, Chappells H, Shove E, 2005 *Infrastructures of Consumption: Environmental Innovation in the Utility Industries*. London: Earthscan.

Varady R, Meehan K, McGovern E, 2009 "Charting the emergence of 'global water initiatives' in world water governance", *Physics and Chemistry of the Earth*, 34, 150-155.

Vincent S, Bowers G, (eds) 1988 *James Bay and Northern Quebec – Ten Years After*, Montreal: Recherches amerindiennes au Quebec.

Vinogradov S, 1996 "Transboundary Water Resources in the Former Soviet Union: Between conflict and cooperation", *Natural Resources Journal*, 36.

Volk M, Liersch S, Schmidt G, In Press "Towards the implementation of the European Water Framework Directive: Lessons learned from water quality simulations in an agricultural watershed", *Land Use Policy*.

Ward C, 1997 *Reflected in Water: A crisis in social responsibility*, Cassell.

Ward D.R, 2002 *Water Wars: Drought, flood, folly and the politics of thirst*, Riverhead Books.

Ward R.C, Robinson M, 1989 *Principles of Hydrology*, McGraw Hill, London.

Warren R, Arnell R, Nicholls R, Price J, 2006 "Understanding the regional impacts of climate change", Research report prepared for the Stern Review, Tyndall Centre Working Paper 90, Norwich: Tyndall Centre.

Warren R, Arnell N, Nicholls R, Levy P, Price J, 2006 "Understanding the regional impacts of climate change", Research report prepared for the Stern Review, Tyndall Centre Working Paper 90, Norwich: Tyndall Centre,

Watts Committee, 1985 *Joint Study of Water Metering – Report of the Steering Group*. HM. Stationary Office. London.

White G.F, 1968 *Water and Choice in the Colorado Basin: An Example of Alternatives in Water Management*. National Research Council Committee on Water. No. 1689. Washington, DC: National Academy of Sciences.

WHO, 1990 *The International Water Supply and Sanitation Decade: end of decade review*, WHO.

Windgassen H, 2006 "Municipal Water Metering – The electromagnetic solution", *World Pumps* (483), 26-29.

Winpenny J, 1994 *Managing Water as an Economic Resource*, London: Routledge.

Wolch J, 1989 "The Shadow State: Transformations in the voluntary sector", in J. Wolch and M. Dear (eds) *The Power of Geography: How territory shapes social life*, Unwin Hyman, pp. 197-220.

Wolf A, 1995 *Hydropolitics Along the Jordan River: Scarce Water and Its Impact on the Arab-Israeli Conflict*, United Nations University.

Wolf A.T, et al., 1999 "International River Basins of the World", *International Journal of Water Resources Development*, 15.

Wolf S, White A, 1995 *Environmental Law*, Cavendish Publishing Limited.

World Bank, 2000 *Natural Resource Management Strategy: Eastern Europe and Central Asia*. World Bank Technical Paper. World Bank, Washington, USA. No. 485, pp. xvi + 142.

World Bank, 1994 *Staff Appraisal Report, Bulgaria: Water Companies Restructuring and Modernisation Project*, 5 May, Report #12317-BUL.

World Bank, 1993 *Water Resources Management*, World Bank Technical Paper. World Bank, Washington, USA, No. 12335.

World Commission on Dams, 2000 *Dams and Development: a new framework for decision-making*, London: Earthscan.

World Health Organisation (WHO), 1997 *Health and Environment in Sustainable Development Five Years After the Earth Summit,* Geneva.

Wouters P, (ed.) 1997 *International Water Law: Selected Writings of Professor Charles B. Bourne,* Kluwer Law International.

Yates C, 2007 "Metered districts, software, help stem water leakage", *Water World*, 23(10), 24-26.

Websites

http://terrassa.pnl.gov:2080/hydrology/biblio.html [Hydrology biblio's ctc]

http://www.soas.ac.uk/Geography/WaterIssues/ [SOAS Water Issues]

http://www.fao.org/waicent/FaoInfo/Agricult/AGL/iptrid/INDEX.HTM [IPTRID]

http://www.usbr.gov/main/programs/international.html [USBR]

http://www.siwi.org/ [Stockholm International Water Institute]

http://www.environment-agency.gov.uk/modules/homepages/MOD31/articlelist english.html [Demand Management Bulletin]

http://www.elsevier.nl/inca/publications/store/6/0/0/6/5/6/index.htt [Water Policy – a Journal]

http://nervm.nerdc.ufl.edu/~mslwww/water.html [Selected sites]

http://www.lboro.ac.uk/well/ [Well-health]

http://www.pbs.org/kteh/cadillacdesert/home.html [Cadillac Desert]

http://www.nwl.ac.uk/ih/ [Institute of Hydrology]

http://www.silsoe.cranfield.ac.uk/ukia/index.htm [UK Irrigation Assoc – Silsoe]

http://www.gwpforum.org/ [Global Water Partnership]

Index